water

GLOBAL CHALLENGES & POLICY OF FRESHWATER USE

Water: Global Challenges & Policy of Freshwater Use

© 2014 National Geographic Learning, Cengage Learning

Photographic Credit: cover
© Michael Yamashita/National Geographic Image Collection

For product information and technology assistance, contact us at
Cengage Learning Customer & Sales Support, 888-915-3276

For permission to use material from this text or product,
submit all requests online at **www.cengage.com/permissions.**
Further permissions questions can be emailed to
permissionrequest@cengage.com.

ISBN: 978-12851-94455

National Geographic Learning
1880 Oak Avenue, Suite 300
Evanston, IL 60201
USA

Cengage Learning products are represented in Canada by
Nelson Education, Ltd.

Visit National Geographic Learning online at **NGL.cengage.com**
Visit our corporate website at **www.cengage.com.**

Printed in the USA.

16 17 18 19 20 21 22

10 9 8 7 6 5 4 3 2

Table *of* Contents

About the Series

Cengage Learning and National Geographic Learning are proud to present the *National Geographic Learning Reader Series*. This groundbreaking series is brought to you through an exclusive partnership with the National Geographic Society, an organization that represents a tradition of amazing stories, exceptional research, first-hand accounts of exploration, rich content, and authentic materials.

The series brings learning to life by featuring compelling images, media, and text from National Geographic. Through this engaging content, students develop a clearer understanding of the world around them. Published in a variety of subject areas, the *National Geographic Learning Reader Series* connects key topics in each discipline to authentic examples and can be used in conjunction with most standard texts or online materials available for your courses.

How the reader works

Each article is focused on one topic relevant to the discipline. The introduction provides context to orient students and focus questions that suggest ideas to think about while reading the selection. Rich photography, compelling images, and pertinent maps are amply used to further enhance understanding of the selections. The chapter culminating section includes discussion questions to stimulate both in-class discussion and out-of-class work.

An eBook will accompany each reader and will provide access to the text online with a media library that may include images, videos, and other content specific to each individual discipline.

Few organizations present this world, its people, places, and precious resources in a more compelling way than National Geographic. Through this reader series we honor the mission and tradition of National Geographic Society: to inspire people to care about the planet.

Access to fresh, clean water is one of the most limiting resources on the planet. Both human civilizations and wildlife depend upon water to survive. Despite the Earth's designation as the "Water Planet," freshwater makes up less than 3 percent of the total planetary water reserve—much of which is unavailable for human use due to its location. Access to freshwater has historically determined where civilizations developed and changes in its abundance or quality have often led to their declines. Today, the lack of adequate water supplies negatively affects the lives of more people throughout the world than perhaps any time in history. Growing populations, changing climates, increases in pollution levels and redistribution of water from historic routes has compounded what has always been a significant global issue. At the same time, advancing technologies and greater communication between affected communities has created a greater capacity to deal with these issues than ever before. Decisions governing this important resource affect people and wildlife in every corner of the planet. Based on the significance of this issue and its impact, global water policy is an increasingly important topic for today's students.

This National Geographic reader explores the complex real world challenges that both individuals and governments face in deciding how to manage global freshwater resources. The selected articles document current conditions and situations throughout the world to illuminate a variety of freshwater policy problems. They contain specific examples of the effects of changing climates on precipitation patterns and explain how growing populations and competing industry interests must adapt to these changing patterns. They examine the controversy over what inherent and legal rights nature and wildlife should have in water policy decisions. Other crucial issues are discussed including the impact of large-scale water pollution and aquifer depletion, restoration potentials and dam removal, freshwater production and distribution technologies, and how international relations influence water usage and rights.

The reader is organized into three units. The first unit gives students a background in freshwater characteristics and importance and provides historical context for current conditions. The second unit examines some of the most critical challenges facing individuals and governments and explores their attempts to address these challenges. The unit takes a case studies approach to highlight the consequences of mismanagement and overuse of existing freshwater resources. The third unit provides additional examples of conflicts that have traditionally surrounded freshwater distribution issues and offers potential solutions that water policy negotiations and agreements can provide. The goal of this final unit is to focus attention on the implications of bringing different interest groups together to meet the long-term goal of sustainable freshwater stewardship for the planet. Collectively, these articles help students from a wide variety of disciplines obtain an appreciation for the importance of water to our survival and the challenges involved in global freshwater policy.

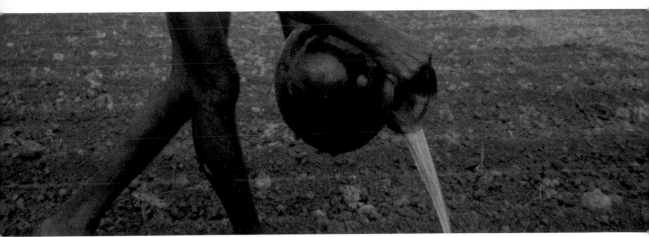

FRESH WATER

Fresh Water discusses the role that water plays in our environment. It gives a brief introduction to the importance that water has on the lives of people and the ecology of various regions and explains how potential changes in the distribution of water may affect population centers and ecosystems. It also discusses regional challenges and potential policy solutions.

When reading this article, you should focus on:

- Why freshwater is important to the distribution and health of wildlife populations,
- Where the primary reservoirs of freshwater are found on the planet and how much is contained in each,
- How potential climate change may alter rain distribution patterns,
- What rights nature has, and
- What is society's role in protecting existing freshwater sources for future populations and dependent wildlife?

Impalas quench their thirst at a water hole (Botswana)

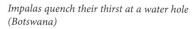

FRESH WATER

By Barbara Kingsolver

Iguacu Falls as seen from the Brazilian observation deck. Iguacu Falls stretches for two miles near the borders of Argentina, Paraguay, and Brazil.

THE AMOUNT OF MOISTURE ON EARTH HAS NOT CHANGED. THE WATER THE DINOSAURS DRANK MILLIONS OF YEARS AGO IS THE SAME WATER THAT FALLS AS RAIN TODAY. BUT WILL THERE BE ENOUGH FOR A MORE CROWDED WORLD?

We keep an eye out for wonders, my daughter and I, every morning as we walk down our farm lane to meet the school bus. And wherever we find them, they reflect the magic of water: a spider web drooping with dew like a rhinestone necklace. A rain-colored heron rising from the creek bank. One astonishing morning, we had a visitation of frogs. Dozens of them hurtled up from the grass ahead of our feet, launching themselves, white-bellied, in bouncing arcs, as if we'd been caught in a downpour of amphibians. It seemed to mark the dawning of some new aqueous age. On another day we met a snapping turtle in his primordial olive drab armor. Normally this is a pond-locked creature, but some murky ambition had moved him onto our gravel lane, using the rainy week as a passport from our farm to somewhere else.

The little, nameless creek tumbling through our hollow holds us in thrall. Before we came to southern Appalachia, we lived for years

Water is life. It's the briny broth of our origins, the pounding circulatory system of the world. We stake our civilizations on the coasts and mighty rivers. Our deepest dread is the threat of having too little—or too much.

in Arizona, where a permanent runnel of that size would merit a nature preserve. In the Grand Canyon State, every license plate reminded us that water changes the face of the land, splitting open rock desert like a peach, leaving mile-deep gashes of infinite hue. Cities there function like space stations, importing every ounce of fresh water from distant rivers or fossil aquifers. But such is the human inclination to take water as a birthright that public fountains still may bubble in Arizona's town squares and farmers there raise thirsty crops. Retirees from rainier climes irrigate green lawns that impersonate the grasslands they left behind. The truth encroaches on all the fantasies, though, when desert residents wait months between rains, watching cacti tighten their belts and *(Continued on page 8)*

Adapted from "Water is Life" by Barbara Kingsolver: National Geographic Magazine, April 2010.

Red lechwe herd foraging on floodplains in the Bengweule swamps (Luangwa Valley, Zambia).

During a 1972 drought in Bangladesh, a farmer dispensed precious water plant by plant. In Iceland the bountiful Kolgríma River inscribes the earth on its seaward path.

(*Continued from page 5*) roadrunners skirmish over precious beads from a dripping garden faucet. Water is life. It's the briny broth of our origins, the pounding circulatory system of the world, a precarious molecular edge on which we survive. It makes up two-thirds of our bodies, just like the map of the world; our vital fluids are saline, like the ocean. The apple doesn't fall far from the tree.

Even while we take Mother Water for granted, humans understand in our bones that she is the boss. We stake our civilizations on the coasts and mighty rivers. Our deepest dread is the threat of having too little moisture—or too much. We've lately raised the Earth's average temperature by .74°C (1.3°F), a number that sounds inconsequential. But these words do not: flood, drought, hurricane, rising sea levels, bursting levees. Water is the visible face of climate and, therefore, climate change. Shifting rain patterns flood some regions and dry up others as nature demonstrates a grave physics lesson: Hot air holds more water molecules than cold.

The results are in plain sight along pummeled coasts from Louisiana to the Philippines as superwarmed air above the ocean brews superstorms, the likes of which we have never known. In arid places the same physics amplify evaporation and drought, visible in the dust-dry farms of the Murray-Darling River Basin in Australia. On top of the Himalaya, glaciers whose meltwater sustains vast populations are dwindling. The snapping turtle I met on my lane may have been looking for higher ground. Last summer brought us a string of floods that left tomatoes blighted on the vine and our farmers needing disaster relief for the third consecutive year. The past decade has brought us more extreme storms than ever before, of the kind that dump many inches in a day, laying down crops and utility poles and great sodden oaks whose roots cannot find purchase in the saturated ground. The word "disaster" seems to mock us. After enough repetitions of shocking weather, we can't remain indefinitely shocked.

How can the world shift beneath our feet? All we know is founded on its rhythms: Water will flow from the snowcapped mountains, rain and sun will arrive in their proper seasons. Humans first formed our tongues around language, surely, for the purpose of explaining these constants to our children. What should we tell them now? That "reliable" has been rained out, or died of thirst? When the Earth

We have been slow to give up on the myth of Earth's infinite generosity. Rather grandly, we have overdrawn our accounts.

seems to raise its own voice to the pitch of a gale, have we the ears to listen?

A world away from my damp hollow, the Bajo Piura Valley is a great bowl of the driest Holocene sands I've ever gotten in my shoes. Stretching from coastal, northwestern Peru into southern Ecuador, the 14,000-square-mile Piura Desert is home to many endemic forms of thorny life. Profiles of this eco-region describe it as dry to drier, and Bajo Piura on its southern edge is what anyone would call driest. Between January and March it might get close to an inch of rain, depending on the whims of El Niño, my driver explained as we bumped over the dry bed of the Río Piura, "but in some years, nothing at all." For hours we passed through white-crusted fields ruined by years of irrigation and then into eye-burning valleys beyond the limits of endurance for anything but sparse stands of the deep-rooted Prosopis pallida, arguably nature's most arid-adapted tree. And remarkably, some scattered families of Homo sapiens.

They are economic refugees, looking for land that costs nothing. In Bajo Piura they find it, although living there has other costs, and fragile drylands pay their own price too, as people exacerbate desertification by cutting anything living for firewood. What brought me there, as a journalist, was an innovative reforestation project. Peruvian conservationists, partnered with the NGO Heifer International, were guiding the population into herding goats, which eat the protein-rich pods of the native mesquite and disperse its seeds over the desert. In the shade of a stick shelter, a young mother set her dented pot on a dung-fed fire and showed how she curdles goat's milk into white cheese. But milking goats is hard to work into her schedule when she, and every other woman she knows, must walk about eight hours a day to collect water. (Continued on page 12)

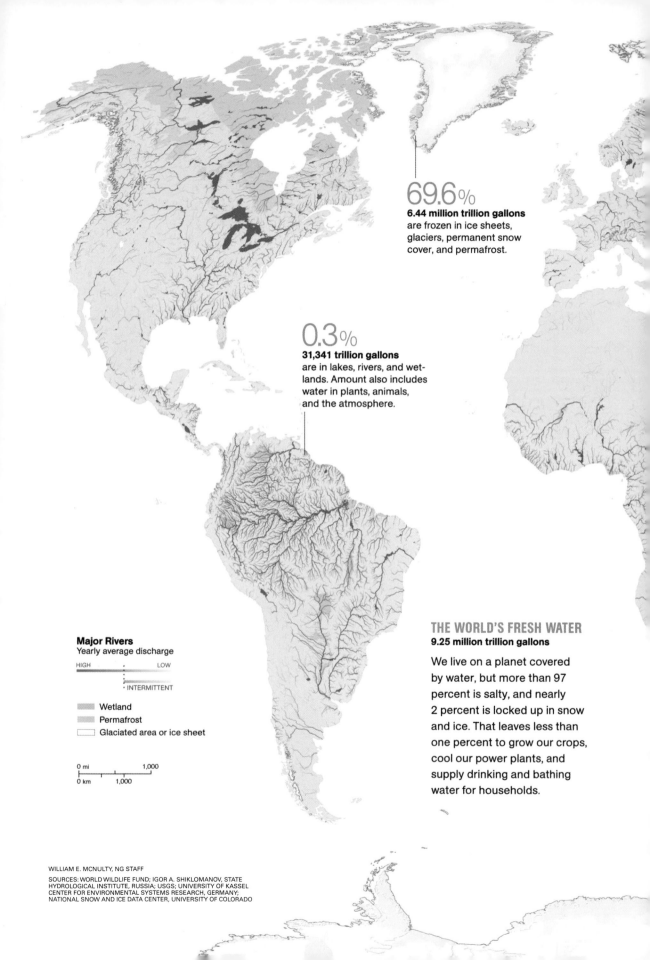

69.6%
6.44 million trillion gallons
are frozen in ice sheets,
glaciers, permanent snow
cover, and permafrost.

0.3%
31,341 trillion gallons
are in lakes, rivers, and wetlands. Amount also includes
water in plants, animals,
and the atmosphere.

Major Rivers
Yearly average discharge

HIGH · LOW
· INTERMITTENT

Wetland
Permafrost
Glaciated area or ice sheet

0 mi   1,000
0 km   1,000

THE WORLD'S FRESH WATER
9.25 million trillion gallons

We live on a planet covered
by water, but more than 97
percent is salty, and nearly
2 percent is locked up in snow
and ice. That leaves less than
one percent to grow our crops,
cool our power plants, and
supply drinking and bathing
water for households.

WILLIAM E. MCNULTY, NG STAFF

SOURCES: WORLD WILDLIFE FUND; IGOR A. SHIKLOMANOV, STATE
HYDROLOGICAL INSTITUTE, RUSSIA; USGS; UNIVERSITY OF KASSEL
CENTER FOR ENVIRONMENTAL SYSTEMS RESEARCH, GERMANY;
NATIONAL SNOW AND ICE DATA CENTER, UNIVERSITY OF COLORADO

30.1%

2.78 million trillion gallons
are beneath the ground
in soil and aquifers fed
by surface seepage.

Groundwater
Average rate of recharge

HIGH MEDIUM LOW

ON NEARLY EVERY CONTINENT,
GROUNDWATER IN AQUIFERS IS BEING
DRAINED FASTER THAN THE NATURAL
RATE OF RECHARGE.

Spawning salmon dominate traffic in the Ozernaya River (Kamchatka, Russia).

(Continued from page 9) Their husbands were digging a well nearby. They worked with hand trowels, a plywood form for lining the shaft with concrete, inch by inch, and a sturdy hand-built crank for lowering a man to the bottom and sending up buckets of sand. A dozen hopeful men in stained straw hats stood back to let me inspect their work, which so far had yielded only a mountain of exhumed sand, dry as dust. I looked down that black hole, then turned and climbed the sand mound to hide my unprofessional tears. I could not fathom this kind of perseverance and wondered how long these beleaguered people would last before they'd had enough of their water woes and moved somewhere else.

Five years later they are still bringing up dry sand, scratching out their fate as a microcosm of life on this planet. There is nowhere else. Forty percent of the households in sub-Saharan Africa are more than a half hour from the nearest water, and that distance is growing. Australian farmers can't follow the rainfall patterns that have shifted south to fall on the sea. A salmon that runs into a dam when homing in on her natal stream cannot make other plans. Together we dig in, for all we're worth.

Since childhood I've heard it's possible to look up from the bottom of a well and see stars, even in daylight. Aristotle wrote about this, and so did Charles Dickens. On many a dark night the vision of that round slip of sky with stars has comforted me. Here's the only problem: It's not true. Western civilization was in no great hurry to give up this folklore; astronomers believed it for centuries, but a few of them eventually thought to test it and had their illusions dashed by simple observation.

Civilization has been similarly slow to give up on our myth of the Earth's infinite generosity. Declining to look for evidence to the contrary, we just knew it was there. We pumped aquifers and diverted rivers, trusting the twin lucky stars of unrestrained human expansion and endless supply. Now water tables plummet in countries harboring half

Karuk tribesmen fish for salmon using a dipnet at Ishi Pishi Falls (Sommes Bar, California).
© 2010/DAVID MCLAIN/National Geographic Image Collection

the world's population. Rather grandly, we have overdrawn our accounts.

In 1968 the ecologist Garrett Hardin wrote a paper called "The Tragedy of the Commons," required reading for biology students ever since. It addresses the problems that can be solved only by "a change in human values or ideas of morality" in situations where rational pursuit of individual self-interest leads to collective ruin. Cattle farmers who share a common pasture, for example, will increase their herds one by one until they destroy the pasture by overgrazing. Agreeing to self-imposed limits instead, unthinkable at first, will become the right thing to do. While our laws imply that morality is fixed, Hardin made the point that "the morality of an act is a function of the state of the system at the time it is performed." Surely it was no sin, once upon a time, to shoot and make pies of passenger pigeons.

Water is the ultimate commons. Watercourses once seemed as boundless as those pigeons that darkened the sky overhead, and the notion of protecting water was as silly as bottling it. But rules change. Time and again, from New Mexico's antique irrigation codes to the UN Convention on International Watercourses, communities have studied water systems and redefined wise use. Now Ecuador has become the first nation on Earth to put the rights of nature in its constitution so that rivers and forests are not simply property but maintain their own right to flourish. Under these laws a citizen might file suit on behalf of an injured watershed, recognizing that its health is crucial to the common good. Other nations may follow Ecuador's lead. Just as legal systems once reeled to comprehend women or former slaves as fully entitled, law schools in the U.S. are now *(Continued on page 16)*

A Baptist minister baptizes a church member in river (Mississippi River, near Natchez, Mississippi).

In the Salmon River, locals prepare to play a game using a greased watermelon (California).

(Continued from page 13) reforming their curricula with an eye to understanding and acknowledging nature's rights.

On my desk, a glass of water has caught the afternoon light, and I'm still looking for wonders. Who owns this water? How can I call it mine when its fate is to run through rivers and living bodies, so many already and so many more to come? It is an ancient, dazzling relic, temporarily quarantined here in my glass, waiting to return to its kind, waiting to move a mountain. It is the gold standard of biological currency, and the good news is that we can conserve it in countless ways. Also, unlike petroleum, water will always be with us. Our trust in Earth's infinite generosity was half right, as every raindrop will run to the ocean, and the ocean will rise into the firmament. And half wrong, because we are not important to water. It's the other way around. Our task is to work out reasonable ways to survive inside its boundaries. We'd be wise to fix our sights on some new stars. The gentle nudge of evidence, the guidance of science, and a heart for protecting the commons: These are the tools of a new century. Taking a wide-eyed look at a watery planet is our way of knowing the stakes, the better to know our place.

Discussion Questions:

- The article states "Water is the ultimate commons." What is meant by this statement in the context of Hardin's "Tragedy of the Commons," and how has this contributed to many of the freshwater problems that we currently face?

- What is meant by the "myth of Earth's infinite generosity" with respect to water? How does the author rationalize that this statement is half right and half wrong?

- How does distance from a reliable source of fresh, clean water affect life in drought-stricken communities?

- What is desertification? How can humans exacerbate desertification?

- How do you feel about Ecuador's move to put the rights of nature in its constitution so that rivers and forests have their own rights? What are some potential positive and negative consequences of this policy? Do you think that other countries should follow suit?

Map Literacy

Use the map contained in this article to explain the following:

- If water covers approximately two thirds of the Earth's surface, why are we in any danger of running out of freshwater?

- Why might certain areas with rapidly expanding populations like China and North Africa be concerned about relying too heavily on pumping water from groundwater aquifers?

OUTLOOK: EXTREME

This article looks at important background science associated with the hydrosphere and potential consequences of a changing climate on rainfall totals, locations and human populations. It also uses a historical context to describe how droughts have affected powerful civilizations in the past.

When reading this article, you should focus on:

- How changes in freshwater distribution may have impacted past civilizations,
- How warming temperatures may affect air circulation patterns, which influence rainfall totals and distributions,
- Why the form that precipitation comes in is so important for many areas of the world, and
- How these changes may affect political stability and regional conflicts.

An Indian tailor in Porbandar saves his antique sewing machine from a monsoon flood.

OUTLOOK: EXTREME

By Elizabeth Kolbert

Desiccated carcass of a zebu cow lies on dried mud (Pantanal, Brazil).

AS THE PLANET WARMS, LOOK FOR MORE FLOODS
WHERE IT'S ALREADY WET
AND DEEPER DROUGHT
WHERE WATER IS SCARCE.

Wet areas are going to get **wetter, and** dry areas **drier.**

The world's first empire, known as Akkad, was founded some 4,300 years ago, between the Tigris and the Euphrates Rivers. The empire was ruled from a city—also known as Akkad—that is believed to have lain just south of modern-day Baghdad, and its influence extended north into what is now Syria, west into Anatolia, and east into Iran. The Akkadians were well organized and well armed and, as a result, also wealthy: Texts from the time testify to the riches, from rare woods to precious metals, that poured into the capital from faraway lands.

Then, about a century after it was founded, the Akkad empire suddenly collapsed. During one three-year period four men in succession briefly claimed to be emperor. "Who was king? Who was not king?" a register known as the Sumerian King List asks.

For many years, scholars blamed the empire's fall on politics. But about a decade ago, climate scientists examining records from lake bottoms and the ocean floor discovered that right around the time that the empire disintegrated, rainfall in the region dropped dramatically. It is now believed that Akkad's collapse was caused by a devastating drought. Other civilizations whose demise has recently been linked to shifts in rainfall include the Old Kingdom of Egypt, which fell right around the same time as Akkad; the Tiwanacu civilization, which thrived near Lake Titicaca, in the Andes, for more than a millennium before its fields were abandoned around A.D. 1100; and the Classic Maya civilization, which collapsed at the height of its development, around A.D. 800.

The rainfall changes that devastated these early civilizations long predate industrialization; they were triggered by naturally occurring climate shifts whose causes remain uncertain. By contrast, climate change brought about by increasing greenhouse gas concentrations is our own doing. It, too, will influence precipitation patterns, in ways that, though not always easy to predict, could prove equally damaging.

Warm air holds more water vapor—itself a greenhouse gas—so a hotter world is a world where the atmosphere (Continued on page 24)

Adapted from "Changing Rains" by Elizabeth Kolbert: National Geographic Magazine, April 2009.

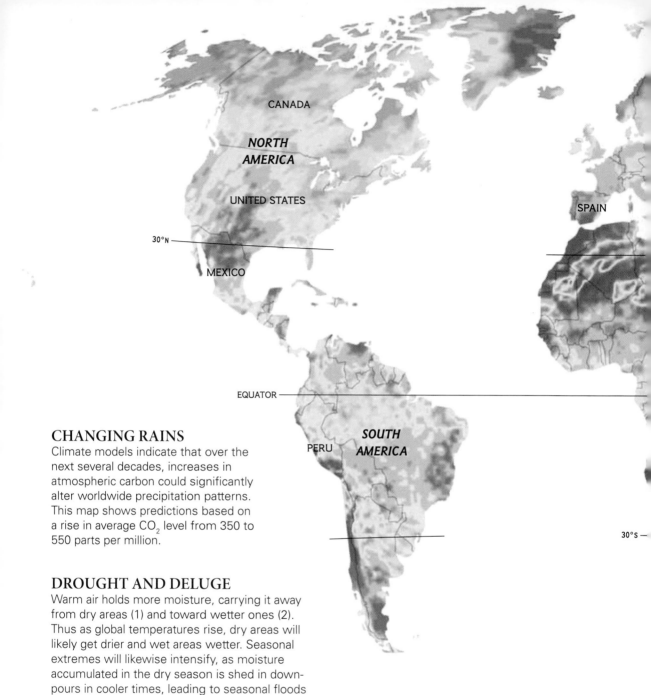

CANADA

*NORTH
AMERICA*

UNITED STATES

SPAIN

30°N

MEXICO

EQUATOR

CHANGING RAINS
Climate models indicate that over the
next several decades, increases in
atmospheric carbon could significantly
alter worldwide precipitation patterns.
This map shows predictions based on
a rise in average CO_2 level from 350 to
550 parts per million.

PERU

*SOUTH
AMERICA*

30°S

DROUGHT AND DELUGE
Warm air holds more moisture, carrying it away
from dry areas (1) and toward wetter ones (2).
Thus as global temperatures rise, dry areas will
likely get drier and wet areas wetter. Seasonal
extremes will likewise intensify, as moisture
accumulated in the dry season is shed in downpours in cooler times, leading to seasonal floods
in regions otherwise prone to drought.

SPREADING DESERTS
Atmospheric warming is also predicted to affect
rainfall by altering global air circulation. At present,
warm air carried from the tropics by circulation
loops called Hadley cells meets cool polar air carried by Ferrel cells in zones around 30° north and
south, creating arid zones. As the planet warms,
these zones are expected to expand and shift
toward the Poles.

SEAN MCNAUGHTON AND LISA R. RITTER (MAP), HIRAM HENRIQUEZ (GRAPHICS),
ALL NG STAFF
SOURCE: GEOPHYSICAL FLUID DYNAMICS LABORATORY, NOAA

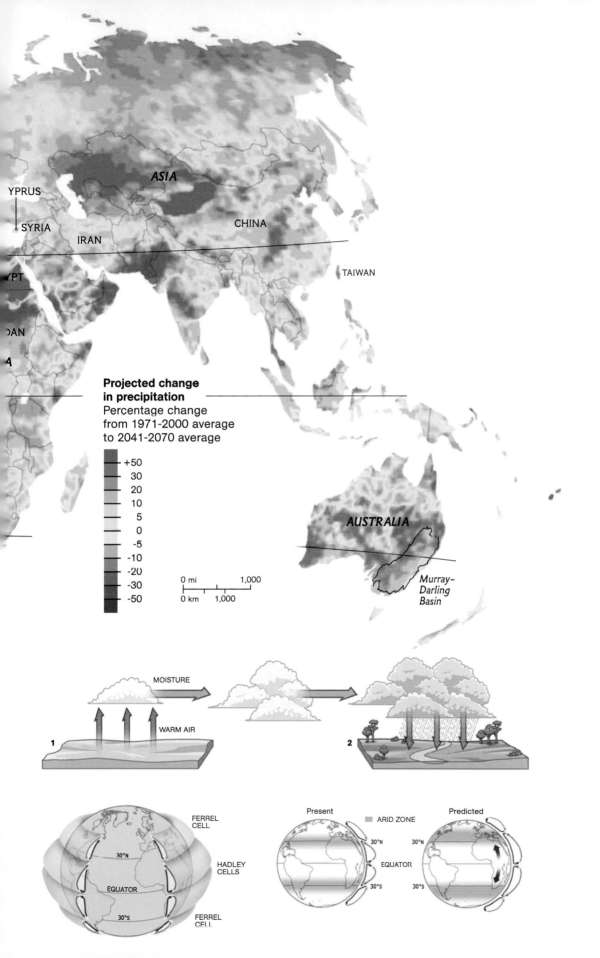

YPRUS

SYRIA

IRAN

ASIA

CHINA

TAIWAN

YPT

AN

**Projected change
in precipitation**
Percentage change
from 1971-2000 average
to 2041-2070 average

+50
30
20
10
5
0
-5
-10
-20
-30
-50

0 mi   1,000
0 km   1,000

AUSTRALIA

Murray–
Darling
Basin

MOISTURE

WARM AIR

1

2

FERREL
CELL

30°N

HADLEY
CELLS

EQUATOR

30°S

FERREL
CELL

Present

ARID ZONE

30°N

EQUATOR

30°S

Predicted

30°N

EQUATOR

30°S

Flooded crops from a broken Mississippi River levee (Oakville, Iowa).
© 2009/JOHN STANMEYER LLC / National Geographic Image Collection

(Continued from page 21) contains more moisture. (For every degree Celsius that air temperatures increase, a given amount of air near the surface holds roughly 7 percent more water vapor.) This will not necessarily translate into more rain—in fact, most scientists believe that total precipitation will increase only modestly—but it is likely to translate into changes in where the rain falls. It will amplify the basic dynamics that govern rainfall: In certain parts of the world, moist air tends to rise, and in others, the moisture tends to drop out as rain and snow.

"The basic argument would be that the transfers of water are going to get bigger," explains Isaac Held, a scientist at the National Oceanic and Atmospheric Administration's Geophysical Fluid Dynamics Laboratory at Princeton University. Climate models generally agree that over the coming century, the polar and subpolar regions will receive more precipitation, and the subtropics—the area between the tropical and temperate zones—will receive less. On a regional scale, the models disagree about some trends. But there is a consensus that the Mediterranean Basin will become more arid. So, too, will Mexico, the southwestern United States, South Africa, and southern Australia. Canada and northern Europe, for their part, will grow damper.

A good general rule of thumb, Held says, is that "wet areas are going to get wetter, and dry areas drier." Since higher temperatures lead to increased evaporation, even areas that continue to receive the same amount of overall precipitation will become more prone to drought. This poses a particular risk for regions that already subsist on minimal rainfall or that depend on rain-fed agriculture.

"If you look at Africa, only about 6 percent of its cropland is irrigated," notes Sandra Postel, an expert on freshwater resources and director of the Global Water Policy Project. "So it's a very vulnerable region."

Meanwhile, when rain does come, it will likely arrive in more intense bursts, increasing the risk of flooding—even in areas that are drying out. A recent report by the United Nations' Intergovernmental Panel on Climate Change (IPCC) notes that "heavy precipitation events are projected to become more frequent" and that an increase in such events is probably already contributing to disaster. In the single decade between 1996 and 2005 there were twice as many inland flood catastrophes as in the three decades between 1950 and 1980.

"It happens not just spatially, but also in time," says Brian Soden, a professor of marine and atmospheric science at the University of Miami. "And so the dry periods become drier, and the wet periods become wetter."

Rain is what scientists call a "noisy" phenomenon, meaning that there is a great deal of natural variability from year to year.

Quantifying the effects of global warming on rainfall patterns is challenging. Rain is what scientists call a "noisy" phenomenon, meaning that there is a great deal of natural variability from year to year. Experts say that it may not be until the middle of this century that some long-term changes in precipitation emerge from the background clatter of year-to-year fluctuations. But others are already discernible. Between 1925 and 1999, the area between 40 and 70 degrees north latitude grew rainier, while the area between zero and 30 degrees north grew drier. In keeping with this broad trend, northern Europe seems to be growing wetter, while the southern part of the continent grows more arid. The Spanish Environment Ministry has estimated that, owing to the combined effects of climate change and poor land-use practices, fully a third of the country is at risk of desertification. Meanwhile, the island of Cyprus has become so parched that in the summer of 2008, with its reservoir levels at just 7 percent, it was forced to start shipping in water from Greece.

"I worry," says Cyprus's environment commissioner, Charalambos Theopemptou. "The IPCC is talking about a 20 or 30 percent reduction of rainfall in this area, which means that the problem is here to stay. And this combined with higher temperatures—I think it is going to make life very hard in the whole of the Mediterranean."

Other problems could follow from changes not so much in the amount of precipitation as in the type. It is estimated that more than a billion people—about a sixth of the world's population—live in regions whose water supply depends, at least in part, on runoff from glaciers or seasonal snowmelt. As the world warms, more precipitation will fall as rain and less as snow, so this storage system may break down. The Peruvian city of Cusco, for instance, relies in part on runoff from the glaciers of the Quelccaya ice cap to provide water in summer. In recent years, as the glaciers have receded owing to rising temperatures, Cusco has periodically had to resort to water rationing.

Several recent reports, including a National Intelligence Assessment prepared for American policymakers in 2008, predict that over the next few decades, climate change will emerge as a significant source of political instability. (It was no coincidence, perhaps, that the drought-parched Akkad empire was governed in the end by a flurry of teetering monarchies.) Water shortages in particular are likely to create or exacerbate international tensions. "In some areas of the Middle East, tensions over water already exist," notes a study prepared by a panel of retired U.S. military officials. Rising temperatures may already be swelling the ranks of international refugees—"Climate change is today one of the main drivers of forced displacement," the United Nations High Commissioner for Refugees, António Guterres, has said—and

contributing to armed clashes. Some experts see a connection between the fighting in Darfur, which has claimed an estimated 300,000 lives, and changes in rainfall in the region, bringing nomadic herders into conflict with farmers.

Will the rainfall changes of the future affect societies as severely as some of the changes of the past? The American Southwest, to look at one example, has historically been prone to droughts severe enough to wipe out—or at least disperse—local populations. (It is believed that one such megadrought at the end of the 13th century contributed to the demise of the Anasazi civilization, centered in what currently is the Four Corners.) Nowadays, of course, water-management techniques are a good deal more sophisticated than they once were, and the Southwest is supported by what Richard Seager, an expert on the climatic history of the region, calls "plumbing on a continental scale." Just how vulnerable is it to the aridity likely to result from global warming?

"We do not know, because we have not been at this point before," Seager observes. "But as man changes the climate, we may be about to find out."

Discussion Questions

How might atmospheric warming influence rainfall in terms of both quantity and distribution? Which areas do we expect to be most affected by these changes and how might they change?

- Why is it problematic to use a single year's data to make generalizations about changing climates? How might natural variability change with a changing climate?

- Why do you think that so many policy makers are concerned about the form water comes in? For example, if regional average input totals of freshwater stay the same, what difference does it make that more is coming in the form of rain and less as snow?

- What role, if any, do you think changing freshwater distributions may play in terms of regional political stability and international relations? What areas do you think may be influenced the most?

Map Literacy

Use the map contained in this article to explain the following:

- How do climate models suggest worldwide precipitation patterns may change as a result of increasing levels of atmospheric carbon dioxide? Which areas do you think may have the hardest time dealing with this?

- Explain what may happen if warming atmospheric temperatures cause descending air from Hadley cells to shift toward the Poles. Why would this shift affect regional freshwater totals?

AUSTRALIA'S DRY RUN

Australia's Dry Run is a geographically specific case study that builds on the information presented in the previous two articles. It illustrates the human side of changing water distributions by telling the stories of a dairy farmer, citrus farmer and fisherman who are all struggling as a result of the drought that is unfolding in the Murray-Darling region of Australia. It illustrates how residents now live with water shortages as a normal part of life and have started to adapt through changing their water use behaviors and sometimes making painful choices.

When reading this article, you should focus on:

- Why settlers originally started farming in the region,
- How their initial actions and subsequent policy decisions have contributed to some of the problems that they currently face,
- How water shortages can pit one resident or business against others in a region based on water allocations,
- How the ecology of coastal lakes that depend on freshwater input may change when water is diverted, and
- What are some future options for cities and rural residents in these areas?

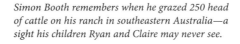

Simon Booth remembers when he grazed 250 head of cattle on his ranch in southeastern Australia—a sight his children Ryan and Claire may never see.

AUSTRALIA'S
DRY RUN

By Robert Draper

Photographs by Amy Toensing

Stressed farmers in the rice-growing town of Coleambally meet to discuss slashed water allocations, which caused a 98 percent drop in rice production from 2006 to 2008. "The meeting was a bit scary," says 74-year-old Frank Whelan (at center), who didn't plant a crop for the first time in more than 50 years.

WHAT WILL HAPPEN WHEN THE CLIMATE STARTS TO CHANGE AND THE RIVERS DRY UP AND A WHOLE WAY OF LIFE COMES TO AN END?
THE PEOPLE OF THE MURRAY-DARLING BASIN ARE FINDING OUT RIGHT NOW.

The climate betrayed him.

On the side of a road somewhere in southeastern Australia sits a man in a motionless pickup truck, considering the many ways in which his world has dried up. The two most obvious ways are in plain view. Just beyond his truck, his dairy cattle graze on the roadside grass. The heifers are all healthy, thank God. But there are only 70 of them. Five years ago, he had nearly 500. The heifers are feeding along a public road—"not strictly legal," the man concedes, but what choice does he have? There is no more grass on the farm he owns. His land is now a desert scrubland where the slightest breeze lifts a hazy wall of dust. He can no longer afford to buy grain, which is evident from the other visible reminder of his plight: the bank balance displayed on the laptop perched on the dashboard of his truck. The man, who has never been rich but also never poor, has piled up hundreds of thousands of dollars in debt. The cows he gazes at through his windshield— that is all the income he has left.

His name is Malcolm Adlington, and for the past 36 of his 52 years he has been a dairy farmer, up at five every morning for the first milking of the day. Not so long ago Adlington used to look forward to a ritual called a dairy farm walk. State agriculture officials would round up local dairy farmers to visit a model farm—often Adlington's, a small but prosperous operation outside of Barham in New South Wales. The farmers would study Adlington's ample grain-fed heifers. They would inquire about his lush hay paddocks— which seeds and fertilizers he favored—and Adlington was only too happy to share information, knowing they would reciprocate when it came their turn. That was the spirit of farming, and of Australia. A man could freely experiment, freely reveal his farming strategies, with the quiet confidence that his toil and ingenuity would win out.

"That," Adlington observes today, "was before the drought came along." A decade ago, Adlington employed five farmhands. "It's just the wife and I now," he says. "The last three years we've had essentially no water. That's what is killing us." *(Continued on page 36)*

Adapted from "Australia's Dry Run" by Robert Draper: National Geographic Magazine, April 2009.

"Sheep graziers are as tough as teak," says Ed Lilburne (at center), who is auctioning off 40,000 animals at a stockyard near the town of Hay. They need to be. Years of drought have forced graziers in the Murray- Darling Basin to sell stock to conserve feed and precious water.

*Blasted by dust and wind, the Booth family farm hasn't seen
normal rainfall since 1991. But Simon Booth isn't giving up: "I
know what this country's capable of with the right type of rain."
Until it falls, he grazes his livestock 400 miles away, hoping
that drought is not, as some say, the "new normal."*

"We're always living in hope," says Malcolm Adlington of Barham, who's sold off his dairy cows to provide for his family. With his government water allocations reduced to near zero, "our farm is for sale," Adlington says, "but nobody's looking at it."

(Continued from page 31) Except there is water. You can see it rippling underneath the main road just a mile from where his truck is parked. It's the Southern Main Canal, an irrigation channel from Australia's legendary Murray River, which along with the Darling River and other waterways is the water source for the South Australia capital of Adelaide and provides 65 percent of all the water used for the country's agriculture. Adlington possesses a license to draw 273 million gallons of water annually from the Murray-Darling River system. The problem is the water has been promised to too many players: the city of Adelaide, the massive corporate farms, the protected wetlands. And so, for the past three years, the New South Wales government has forbidden Adlington from taking little more than a drop. He still has to pay for his allocation of water. He just can't use it. Not until the

drought ends. Adlington finds himself chafing at the unfairness of it all. "It's the lack of rain," he says, "but also the silly man-made rules." Those rules seem to favor everyone except farmers like him. Meanwhile, he's selling off his treasured livestock.

"It's easy to get depressed," he says in a calm, flat voice. "You ask yourself, Why have I done it?" Malcolm Adlington didn't use to doubt himself, but then he has not been himself lately. The drought has depleted more than just his soil. He finds himself bickering with his wife, Marianne, hollering at the kids. He can't afford the gas to take Marianne into town as he used to. With all of the other farmhouses closing up, the nearest boy for his son to play with now lives ten miles away.

Adlington has put his own family acreage up for sale. "Haven't had one person look at it," he says. Not his first choice, obviously. Not what an Adlington would ever wish to do. But when

Flooded by Hume Dam in the 1920s, the Murray's riverbanks were once thick with river red gum trees that captured moisture and helped drive cycles of rainfall. The skeletons of submerged trees are now visible in a ponded portion of the river's upper reaches, exposed by the lowest water levels in decades.

the hell did his dad or granddad ever have to deal with a bloody seven-year drought?

It has been three parched years since any dairy farm walk that Adlington can remember. Instead, there are morale-boosting events with upbeat monikers like Tackling Tough Times or Blokes' Day Out—or Pamper Day, which Adlington's wife happens to be attending today. At Pamper Day, a few dozen farming women receive free massages and pedicures and hairstyling advice. A drought-relief worker serves the women tea and urges them to discuss what's on their minds. They all share different chapters of the same story.

"It's been two years without a crop."

"The family farm is on its knees."

"We sold most of our sheep stock—beautiful animals we'd had for 20 years."

"I can't stand lying in bed every night and hearing the cattle bellow from hunger."

Still, the most poignant gatherings are out of public view. One takes place in a modest farmhouse near Swan Hill. A government rural financial counselor sits at the kitchen table, advising a middle-aged stone-fruit farmer and his wife to declare bankruptcy, since their debt exceeds the value of their farm and a hailstorm has just ravaged their crop.

Holding his wife's hand, tears leaking out of his eyes, the farmer manages to get out the words: "I have absolutely nothing to go on for."

The woman says she checks every couple of hours to make sure her husband is not lying in his orchard with a self-inflicted gunshot wound in his head. When the meeting is over, the counselor adds their names to a suicide watch list.

Back in Barham, Malcolm Adlington sits alone in his truck going *(Continued on page 42)*

RAINFALL
Murray-Darling Patterns
2001-2008

Above average / Average / Below average / Much below average / Lowest

Murray-Darling Anomalies
1910-2007, in mm of rain (base average 1961-1990)

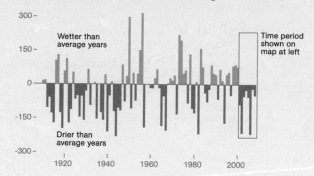

Wetter than average years

Drier than average years

Time period shown on map at left

WARMING
Average Temperature Trends
1970-2007, in degrees Celsius

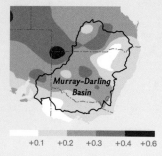

+0.1 +0.2 +0.3 +0.4 +0.6

Anomalies
1910-2007, in degrees Celsius (base average 1961-1990)

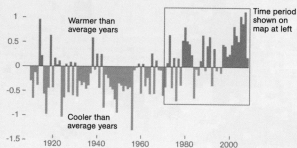

Warmer than average years

Cooler than average years

Time period shown on map at left

AUSTRALIA

Murray-Darling Basin

Goyder's Line

ADI
PIP

Adelai

L
Alexandr

PARCHED LANDS

The dry and marginally fertile lands of the Murray-Darling Basin were transformed into the breadbasket of Australia through a massive water-management program that dammed rivers, filled reservoirs, and tapped water for irrigation and other human needs. It was a precarious balancing act, until seven years of drought—and decades of warmer temperatures—brought farmers to their knees. Stoked by drought, deadly bushfires dealt another blow early this year.

MARTIN GAMACHE, NG STAFF

SOURCES: AUSTRALIAN BUREAU OF METEOROLOGY; AUSTRALIAN BUREAU OF STATISTICS; BRETT BRYAN, CSIRO; GEOSCIENCE AUSTRALIA; MURRAY-DARLING BASIN AUTHORITY

Legend

- ■ Canals and irrigated cropland
- ■ Nonirrigated cropland
- ■ Cattle and sheep grazing
- ■ Nonagricultural land use
- ○ Dairy*
- △ Cotton*
- ◇ Rice*
- □ Vineyards, fruits, and vegetables*

* 2001 CLUSTERS

Charleville

Warrego

Paroo

Condamine

Balonne

DARLING DOWNS

Brisbane

Gold Coast

St. George

Weir

Dumaresq

Macintyre

Source of the Darling

QUEENSLAND

NEW SOUTH WALES

Barwon

Gwydir

Moree

NEW ENGLAND

30°S

Darling

Bourke

Bogan

Macquarie Marshes

Namoi

Tamworth

MURRAY – DARLING BASIN

Lake Menindee

Dubbo

Lachlan

Willandra Cr.

SOUTH AUSTRALIA

RIVERLAND

Darling

Mildura

Hay

Murrumbidgee

Barellan

Cooltong

Renmark

gan

Murray

Coleambally

MALLEE

RIVERINA

Edward

Deniliquin

Wagga Wagga

Sydney

Canberra

Australian Capital Terr.

Swan Hill

Barham

MULWALA CANAL

Lake Mulwala

Albury

Great

The Coorong

Lake Boga

VICTORIA

Murray

Shepparton

Wodonga

Murray

HUME DAM

Snowy Mts.

Australian Alps

Source of the Murray

er

WIMMERA

Bendigo

DARTMOUTH DAM

0 mi 50
0 km 50

AN
AN

Melbourne

145°E

THIRSTIEST CROPS

Agriculture uses up most of the water in the Murray-Darling Basin, though only a small portion of its cropland is irrigated. Its thirstiest crop is cotton, which in 2005-2006 used 20 percent of the basin's water, followed by dairy farming, livestock, and rice. Grapes, other fruits, and vegetables use far less water and generate more income for the region than cotton.

A wet spot is all that's left of Lake Boga, once a water-sports playground, since its tributary was tapped for irrigation in 2007. Fish died by the thousands, and Lake Boga's resort-based economy was left high and dry. "We're a very proud little town," says resident Pauline Coke. "We're devastated."

(Continued from page 37) nowhere—watching his herd dwindle, his meadows receding into desert scrubland. All he can do is watch.

The world's most arid inhabited continent is perilously low on water. Beyond that simple fact, nothing about Australia's water crisis is straightforward. Though Australians have routinely weathered dry spells, the current seven-year drought is the most devastating in the country's 117 years of recorded history. The rain, when it does fall, seems to have a spiteful mind of its own—snubbing the farmlands during winter crop-sowing season, flooding the towns of Queensland, and then spilling out to sea. To many, the erratic precipitation patterns bear the ominous imprint of a human-induced climate shift. Global warming is widely believed to have increased the frequency and severity of natural disasters like this drought. What seems indisputable is that, as Australian environmental scientist Tim Kelly puts it, "we've got a three-quarters of a degree [Celsius] increase in temperature over the past 15 years, and that's driving a lot more evaporation from our water. That's climate change."

It has taken a while for Australia to wake up to that reality. After all, the country was transformed by rough-country optimists unfazed by living on one of the least fertile landscapes on Earth. Australian scientist Tim Flannery calls it a "low-nutrient ecosystem," one whose soil has become old and infertile because it hasn't been stirred up by glaciers within the past million years. The Europeans who descended on the slopes of the Murray-Darling Basin—a vast semiarid plain about the size of Spain and France combined—were lulled by a string of mid-19th-century wet years into thinking they had discovered a latter-day Garden of Eden. Following the habits of their homelands, the settlers felled some 15 billion trees. Unaware of what it would mean to disrupt an established water cycle by uprooting vegetation well adapted to arid conditions, the new Australians introduced sheep, cattle, and water-hungry crops altogether foreign to a desert ecosystem. The endless plowing to encourage Australia's new bounty further degraded its soil.

And so a river became the region's lifeline. Like America's Mississippi River, the 1,600-mile Murray carries mythological significance, symbolizing endless possibility. Its network of billabongs, river red gums, Murray cod, and black swans are as affixed to the Australian ethos as the outback. From its headwaters in the Australian Alps to its destination at the Indian Ocean, the slender river meanders along a northwestern course, fed by the currents of the Murrumbidgee and Darling Rivers as it cuts a long borderline between New South Wales and Victoria before entering the semiarid brush country of South Australia and plunging toward the ocean at Encounter Bay. That its journey appears unhurried, even whimsical, adds to the river's legend.

Progress, for Australians, has involved bending the Murray River to their will. Over the past century, it has been mechanized by an armada of weirs, locks, and barrages, so that the flows will be of maximum benefit to the farmers who depend on irrigation in the Murray-Darling Basin. As a result, says former commonwealth water minister Malcolm Turnbull, "we've got an unnatural environment in the river. Because it's regulated, the river now runs high when nature would run it low, and low when nature would run it high." That manipulation had unintended consequences. Irrigation caused salinity levels to skyrocket, which in turn poisoned wetlands and rendered large stretches of acreage unfit for cultivation.

Such was the rickety state of Australia's water supply even before the drought fell on it like a mallet, delivering a psychic blow for which the plucky land down under was not prepared. The crisis has pitted one state against another, big cities against rural areas, environmental managers against irrigators, and small farms against government-backed superfarms in a high-stakes competition for a shrinking commodity. Well beyond the national breadbasket of the Murray-Darling Basin, every major urban area has faced the clampdown of water restrictions and the subsequent browning of its revered English gardens and cricket ovals. The trauma is particularly acute in rural bastions of self-reliance, like the New South Wales dairy community inhabited by Malcolm Adlington, which are fast becoming ghost towns. Whole crops have been wiped out by heat stress and low moisture, while entire growing sectors—rice, cotton, citrus—face collapse.

The once quintessential Australian swagger has now come to resemble, in the wake of the water crisis, what Swiss psychiatrist Elisabeth Kübler-Ross famously termed the "stages of grief": denial, anger, bargaining, depression, acceptance. In what is shaping up to be a cautionary tale for other developed nations, the world's 15th biggest economy is learning hard lessons about the limits of natural resources in an era of climate change. The upside is that Australians may be the ones to teach those lessons to the rest of the industrialized world.

In the Riverland district of South Australia, a 48-year-old man drives through his citrus orchard on a bulldozer, mowing down 800 of his Valencia and navel orange trees. The man knows what he is doing. Something must give. For decades the mighty Murray River transformed this land into a lush patchwork of olive, citrus, apricot, and avocado orchards. But now the water bureaucrats have announced

> The drought fell on Australia like a mallet, delivering a psychic blow for which the plucky land down under was not prepared.

that South Australians may use only 16 percent of their annual allocation. And so Mick Punturiero, a third-generation farmer of Italian descent, has made a hard choice: He elects to sacrifice his orange trees and reserve what water he has for his prized lime orchard. Underneath the roaring of the engine, Punturiero hears the cracking of muscular trunks he has nurtured for 20 years. And what roils inside him is something darker than sorrow.

A few weeks later two state officials come to Punturiero's village of Cooltong, just outside Renmark, a few hours' drive from Adelaide. They have an announcement to make. The catchment levels at Hume Dam have been revised, and it's good news: The water allocation has been doubled, to 32 percent! The farmers in attendance are not overjoyed. Truthfully, with the drought bearing down on them, 32 percent of what they need is not enough to save their orchards. All Punturiero can think is, I could have kept my orange trees.

Two months later, Punturiero is still possessed of operatic rage as he pours a guest some homemade lime juice and drops his meaty frame into a chair. Why has it taken them so long to recognize this water crisis? he demands. "Let's go to THEIR house! Tell them which child THEY have to sacrifice to save their whole family! Let's put THEIR family in a pile!"

He takes a deep breath. "I get very upset talking about this issue," he says. "I get very, very, VERY agitated over it. End of the day, what's been done is criminal." As to the actual crime and its perpetrators, Mick Punturiero flails with theories. Mostly he blames government officials who encouraged agricultural development beyond sustainable levels. Even in his more reflective moments, he does not entertain the notion that the problem arises

Lettuce grower Donato Gargaro irrigates seedlings with water from the Murrumbidgee River near Hay. About 95 percent water, lettuce is best grown in winter, when rain can augment irrigation and temperatures drop, reducing the water lost to evaporation.

from the folly of growing citrus on the wrong side of "the line."

The line is Goyder's Line, a boundary that marks the limit of sufficient rainfall for crops to grow in South Australia. In 1865 a surveyor named George Goyder set out on a remarkable journey by horseback to trace the point where grassland gave way to sparse bush country. Australia's settlers relied on Goyder's Line to demarcate arable land from land unsuitable for agriculture. Except when they didn't: Renmark, for instance, lay on the wrong side of Goyder's Line, but that did not stop two Canadian brothers named Chaffey from developing an irrigation system in Renmark two decades after the surveyor's warning.

As it turns out, the Chaffeys were three decades ahead of their time. The Australian government inaugurated its first "soldier settlement" scheme after World War I, offering land, water, and farm machinery to veterans. In the decades that followed, orchards and vineyards and wheat fields miraculously sprang up from former scrub desert north of Goyder's Line. Canal after canal was dug to deliver the Murray's water to the new farmland—and later, to sprawling irrigation districts dedicated to the nascent (and highly water-thirsty) rice industry. By the early 1970s, Australia was a major exporter of such crops, its farming lobby had emerged as a formidable political force, and the government was selling off water licenses to any bloke who fancied being his own boss and who wouldn't whinge when the odd drought came along.

Mick Punturiero's grandather was a Calabrian émigré who bought his first acreage from a retiring World War II veteran, one of thousands more soldiers enticed by the government to develop the basin. The audacity of farming in such an arid area was not readily apparent to Punturiero's grandfather, who had no education other than in how to grow an exquisite grape.

Soon the Murray began to run low, and fields started to salt up. Unfortunately, the prescriptions only helped spread the disease. Leakproof irrigation technology meant that less water returned to the system. Salt interceptors kept crops from being poisoned, but only by pumping out limitless quantities of water. In 1995 the Murray-Darling Basin Commission finally introduced a cap on how much water each state could draw from the river. But the binge didn't end. Farmers who owned water rights but had never used them proceeded to sell their now coveted "sleeper licenses" to others who would. Industrialists were offered tax incentives to create superfarms and introduced vast olive and almond groves to the basin.

Meanwhile, the governments of New South Wales and Queensland routinely flouted the extraction cap and continued to hand out licenses. "The increase in diversions from the Murray River in the late nineties was rather like drinkers in a bar," says Malcolm Turnbull. "The barkeeper says, 'Last orders, gentlemen.' And everyone rushes in to drink as much as they can before they get thrown out. That's what we were doing. Just as it became apparent that resources were overtaxed, there were more claims on it."

A decade ago, Mick Punturiero had grown to be South Australia's biggest lime producer and was doing all the right things. He employed the latest water conservation

> **A**delaide may have the dubious distinction of being the first industrialized city to live in a **constant state of water shortage.**

technology. What water he did not need he donated back to the state for environmental usage. Even so, he could see where the increasing demands on the Murray would lead. He recalls warning a state official in the late 1990s, "You need to stop this development. We're poorly managing our water resources."

He remembers the official's words as if uttered yesterday: "Mick, you can't control progress."

Then came the drought, which began like any other, in 2002. But it has not ended, and now the binge is over. Though dryland farmers who depend on rain have watched their corn and wheat fields dwindle into dust plains, they at least have been accustomed to braving parched seasons. By contrast, "irrigated farmers have always had water, and never in their wildest dreams did they think somebody would turn the tap off," says rural financial counselor Don Seward. But as the drought advanced, the allocations have plummeted: 95 percent. Then 50. Then 32. And now, in Mick Punturiero's case, back to 16 percent.

"The river's no different from the highways every Australian pays for through his taxes," he argues. "Every Australian has paid for the locks. We've paid for the Dartmouth Dam, which was supposed to drought-proof South Australia. So why don't you give me my full allocation? Give it to me! It's rightfully mine!"

Punturiero sees himself as the faithful caretaker of land that the Australian government gave to reward the service of young men who died on the sands of Gallipoli. He sees that land as a gold ingot that the government has turned into a lump of lead. He sees powerful interests profiting at his expense. *(Continued on page 48)*

Spin masters, overheated farm boys near Deniliquin put the Mulwala Canal to good use on a cloudless summer day. The largest irrigation canal in the Murray-Darling Basin, the channel transports water from Lake Mulwala more than a hundred miles to the heart of New South Wales farm country.

To water their garden, the Charter family of Hallett Cove—including parents Carl and Leita—shower together and catch the runoff in buckets. It's one of many measures promoted by the state of South Australia that the family has used to cut its water consumption in half.

(Continued from page 45) He sees new irrigators downriver sucking the system dry. He also sees fellow farmers much like his grandfather, who never bothered to put a dime into savings, tumbling into insolvency. Or committing suicide. And he understands their bottomless despair. He feels it himself at times—"boxed into a corner," he says in a suddenly depleted voice, "and I can't defend my family no more."

But fury returns. Anger is all Mick Punturiero has at the moment. He will not go down without a fight—that he pledges: "You won't see me crawling off the farm on me hands and knees—not unless I see some bloody heads roll first!"

I t is hard for many Australians to reconcile the sputtering, surgically disfigured version of the Murray River with the shimmering idyll of their younger days. At the river's mouth, a flourishing ecosystem had long been nourished by the natural ebb and flow of seawater and fresh water. The ocean would rush in when the river ran low and then be pushed out by fresh water as the first hard rains drained down the Murray to the sea. Today the over-allocation of irrigation water, coupled with the drought, has brought the river to a virtual standstill. So that the beleaguered Murray can meet the sea, its mouth must be dredged around the clock. Without dredging, the mouth would silt up, cutting off fresh water to the lagoon ecosystem called The Coorong and to nearby Lake Alexandrina.

It is here, every morning, that a 65-year-old silver-haired fisherman in waders and a

Windbreaker navigates his aluminum boat out into the waters of Lake Alexandrina, or what is left of it. Long humps of silt-covered land rise up out of the water. Since most everyone else in his line of work has moved away, Henry Jones has the lake to himself—not counting the pelicans, though he, in fact, does count them, thinking: Maybe a tenth of what there was. And no white ibis. No blue-billed duck. Edging up to the northern Coorong lagoon, Jones reaches into the water to collect his gill nets. Among his catch there is not a single silver perch or Murray cod or bony bream. The salty water has done them in. Only carp survive. Dozens of carp, which did not even exist in the lower lakes a quarter century ago, and whose presence signals the demise of the freshwater environment.

Jones has adapted to the changes in a way the vanishing species cannot. He has found retailers who will buy all the carp he can catch. And truthfully, he could adapt further. If, as is expected, the government constructs a weir near the bottom of the river to give urban dwellers in Adelaide more water, Lake Alexandrina and its sibling Lake Albert would become saltwater lakes. "Personally, I'd probably be better off catching mullet, flounder, black bream, and a couple of other marine species," he says as he sits at the dining room table of the house he built 40 years ago. "But it's just not right. These lakes have always been freshwater. It's just a massive change. It's nonsense."

The drought has left his community reeling. Local winemakers have recently been informed that the Murray River would no longer be available for their vineyards. And Jones is a close friend to the elders of the Ngarrindjeri Aboriginal people, whose 30,000-year domain over the river abruptly ended when the expedition led by Capt. Charles Sturt

It's up to Australia to show the rest of the world what the new landscape will be—one that's come to terms with limitations.

arrived at the Murray's mouth in 1830. For the Ngarrindjeri, the drought has led to the disappearance of black swan eggs, freshwater mussels, and other sacred totems that are vital to their spiritual and physical nourishment.

Still, in the scramble to claim a share of Australia's diminishing water supply, these people at least have a voice. The creatures of the lakes and wetlands do not. "In a crisis, the entitlement the environment supposedly has is totally subjective to political whims," says Murray River environmental manager Judy Goode, who refers to herself as "the manager of dead and dying things." Even protected ecosystems—such as The Coorong and, in the northern basin, the Macquarie Marshes of bird-nesting legend—receive no special dispensation, so long as there is a "critical human need" to be met.

So Henry Jones has become the de facto voice for the dead and dying, delivering a well-honed, if mournful, monologue to whoever will listen: All the systems are on the point of collapse. Two-thirds of The Coorong is already dead—its salinity is almost that of the Dead Sea. What Jones finds, as he travels around the basin to argue that water must be allocated for his Coorong and his lakes, is a sentiment that the whole water crisis is the environmentalists' fault anyway. The greenies are derided for their shrill sanctimony. Farmers express indignation that any of their precious "working river" is lost to the sea. They tell Jones that it makes more sense to divert the Murray all the way inland, officially consigning the river to eternal servitude as an irrigation channel, while fishermen buck up and learn to live off the sea. In cotton-growing areas wholly dependent on irrigation, Jones says, "I'm lucky to get out with my life."

The Coorong represents only one glaring example of the Murray-Darling Basin's imperiled ecosystem. For example, Australian scientists and government officials were caught unaware when farther upriver some invisible drought-tolerance threshold was crossed and hundreds of thousands of river red gum trees—in the world's biggest such forest—suddenly died. And of late, a fresh concern has emerged: that the wetlands may be brewing toxins. Robbed of their seasonal flushing, and instead unnaturally submerged for decades, the swamps have become so dry that the crusted silt has reacted with air to form large surfaces of sulfuric acid. Scientists haven't fully gauged the threat to animals and people. For now, as University of Adelaide water economist Mike Young observes, "you wouldn't want to put your hand in it."

Adelaide may have the dubious distinction of being the world's first industrialized city to live in a constant state of water shortage. Its unhealthy reliance on the Murray—up to 90 percent of its water supply in low-rainfall periods—is symbolized by two unsightly pipelines that stretch more than 30 miles from the river to the city's water tanks. Since shortly after the drought's onset in 2002, the South Australia capital has been on water restrictions. Its residents dutifully cart buckets of used shower and washing machine water outside to their gardens. Native plants and artificial lawns are de rigueur. The racks of hardware stores are crammed with soil wetters, gray water diverter hoses, water-restricting shower nozzles, four-minute shower timers, and other tributes to water austerity. The radio "talk-back" shows have

But it's just not right. These lakes have always been freshwater. It's just a massive change. It's nonsense.

become reliable outlets for ranting about this or that water abuser.

Still, civic virtue is no substitute for lasting reform. The nation's water crisis won't be solved by "drought-proofing" Adelaide, which, despite its dependence on the Murray, claims only 6 percent of the total drain on the river. "South Australia's very aware that they're living precariously," says Wilderness Society environmental activist Peter Owen. "We're not going to save our river system by standing in buckets."

Meanwhile, outside of the Murray-Darling Basin, the drought has exposed serious flaws in the water resources of Sydney, Melbourne, and Brisbane, among other urban areas. The hard lesson of Australia's dry run is that the country's jaunty boosterism no longer suffices as the way forward. "I work on the assumption that we're going to see more episodes of this type of drought in the future because of climate change," says Malcolm Turnbull, whose Liberal Party leader John Howard, a longtime climate change skeptic, was turned out of office in November 2007. "A prudent minister assumes it's going to get hotter and drier, and plans accordingly."

But what does this mean, really? Will it mean the construction of expensive desalination plants in Adelaide, Sydney, and elsewhere, with escalating energy bills? Will it be possible to develop drought-resistant crop varieties to keep food production up? Or to drastically reduce the water needs of dairy farmers who use a thousand gallons of water for each gallon of milk they produce? Will the Murray River's hard labor continue, or will it see mercy? A robust new landscape is required, and it's up to Australia to show the rest of the industrialized world what that new landscape will be.

"It hurts," says Frank Eddy, to destroy his own healthy peach trees, but the drought and reduced allocations have forced him to cull thousands of older trees on his orchards near Shepparton, Victoria.

For starters, it may be a landscape that's come to terms with limitations. Goyder's Line is even more relevant today, as drought and climate change give new urgency to the question of how intensively marginal agricultural land should be worked—or whether it should be left fallow.

After all, the final stage of coping with loss is acceptance. Back in 1962 Frank Whelan was the third farmer in his New South Wales district to receive a water allocation to grow rice, six years before the town of Coleambally was incorporated. Until this season he always had a crop. Although he's 74, his memory is as clear as his eyes. Droughts, market fluctuations, wrangles with the government, and, yes,

incessant sniping by environmentalists that rice requires enormous quantities of water and therefore has no rightful place on this semi-arid continent—Whelan remembers Coleambally prospering through all the adversity. He remembers town gatherings when the news was almost always good, because the irrigation water was always there.

Today the mood is different as Whelan sits in the local bowling hall with 200 fellow farmers. For four hours they listen as a panel of experts say there will be no irrigation water for Coleambally for the foreseeable future. They are suggesting new economic avenues for the town—things that have nothing to do with rice. *(Continued on page 54)*

Scanning the horizon for rain is a way of life for the Kenny girls, Hannah (at left) and Alice. Their family have been dryland farmers of wheat near Barellan for three generations. "We will have a harvest this year," says the girls' mother, Julie. But just barely—months have passed without a single drop of rain.

(Continued from page 51) A number of farmers voice their outrage. They blame the bureaucrats. They blame the environmentalists. They blame New South Wales. But Whelan says nothing. He just sits there, his pale eyes blinking, occasionally rubbing his wrinkled forehead with a hand that includes two fingers mangled by a farm equipment accident.

He has seen this coming. With the onset of the drought, he compacted his soil with a padfoot roller to minimize leakage. He began to cut off some of his acreage from water. Then still more acreage. All the while, the lifelong farmer watched as national production of rice dropped from more than a million tons a year to 21,000, contributing to the food shortage being felt across the globe. Australia, which has served as a food bowl to the world, is searching for a future. Whatever that future may be, Whelan knows the rice-growing town of Coleambally will never play the same role. And so after the meeting breaks up, a fellow farmer sidles up to him and asks, "Well, what do ya think, mate?"

You won't see me crawling off the farm on me hands and knees—not unless I see some bloody heads roll first!

The question is one that will continue to preoccupy Coleambally for some time to come. At one point, residents actually tossed in the towel and offered to sell the entire town and its water supply to the commonwealth for $2.4 billion. A few days later, they rescinded the offer, digging in their heels and insisting the town will remain a vital food provider.

The wrangle will continue, in Coleambally and throughout Australia. But some have arrived, however reluctantly, at a point of acceptance. A year after the reporting for this story began, dairy farmer Malcolm Adlington sold off the rest of his cattle and now drives a minibus for a living. The citrus grower Mick Punturiero uprooted half of his orchard and acknowledges that he will probably be unable to continue farming. And on this night in Coleambally, Frank Whelan makes a decision as well.

"Oh," he replies to his fellow rice farmer with a sad smile, "I think I'll go home and retire."

Discussion Questions:

- Why have many local farmers and ranchers had to sell off their lands? Is this a result of natural occurrences, water management policy, or both?

- According to the article, why is Australia considered to be a "low nutrient ecosystem"? How do past glaciers have an effect on the productivity of agriculture lands?

- If the Murray-Darling Basin is such a dry area naturally, then why did farmers settle there in the first place? How have their actions subsequently compounded the problem?

- What role should government policy have in shaping water rights? How should governments manage water allotments in drought-prone areas during good times and bad? Should they encourage agricultural development in areas that may not be able to sustain it in the future?

- The article poses a number of potential options that Southeastern Australian cities could look to, including the use of desalinization plants, the development of drought-resistant crops, and the reduction of allocations to existing farmers. What would you do if you were in charge?

Map Literacy

Use the map contained in this article to explain the following:

- Identify where major agricultural products are grown in the region. Are they found in irrigated or non-irrigated areas of the Murray-Darling Basin? What does this imply about their sustainability in a changing climate where droughts (and sometimes floods) are becoming more common?

- How well do rainfall and temperature trends correlate with each other? Explain why you think this occurs.

BITTER WATERS

Bitter Waters describes how industrialization has resulted in large-scale pollution of China's Yellow River. It focuses on specific causes of this widespread pollution, relates it to economic growth and government policies, and follows some of the personal stories of local residents affected by the change in water quality. It also discusses some of the government's technological strategies to increase rainfall and redistribution of water to other areas of the country, and its more recent policy shift towards restoration and greater levels of environmental protections.

When reading this article, you should focus on:

- How the Yellow River has changed and what are the primary causes of these changes,
- How have these changes have affected the local people and levels of political activism,
- Why environmental damage and uncertainty is bad for the Chinese economy,
- How dam construction has led to ecological and environmental threats along the Yellow River, and
- How the new canal system is supposed to relieve the pressure off the Yellow River.

Water fouled by a fertilizer factory in Inner Mongolia steams as it seeps toward the upper reaches of the Yellow River

BITTER WATERS

By Brook Larmer

Photographs by Greg Girard

Beside her father's deathbed, Wu Qiuhua waits. Within hours Wu Qizhi, 60, was dead from cancer of the esophagus. He lived in Wuxin, one of many "cancer villages" along the lower portion of the Yellow River, where pollution has caused a spike in cancer rates. "No matter what we do," says Wu Qiuhua, "it's useless."

A CRISIS IS BREWING

IN CHINA'S NORTHERN HEARTLAND AS ITS LIFELINE, THE YELLOW RIVER, SUCCUMBS TO POLLUTION AND OVERUSE.

I always thought this was the most beautiful place under heaven.

Not a drop of rain has fallen in months, and the only clouds come from sandstorms lashing across the desert. But as the Yellow River bends through the barren landscape of north-central China, a startling vision shimmers on the horizon: emerald green rice fields, acres of yellow sunflowers, lush tracts of corn, wheat, and wolfberry—all flourishing under a merciless sky.

This is no mirage. The vast oasis in northern Ningxia, near the midpoint of the Yellow River's 3,400-mile journey from the Plateau of Tibet to the Bo Hai sea, has survived for more than 2,000 years, ever since the Qin emperor dispatched an army of peasant engineers to build canals and grow crops for soldiers manning the Great Wall. Shen Xuexiang is trying to carry on that tradition today. Lured here three decades ago by the seemingly limitless supply of water, the 55-year-old farmer cultivates cornfields that lie between the ruins of the Great Wall and the silt-laden waters of the Yellow River. From the bank of an irrigation canal, Shen gazes over the green expanse and marvels at the river's power: "I always thought this was the most beautiful place under heaven."

But this earthly paradise is disappearing fast. The proliferation of factories, farms, and cities—all products of China's spectacular economic boom—is sucking the Yellow River dry. What water remains is being poisoned. From the canal bank, Shen points to another surreal flash of color: blood-red chemical waste gushing from a drainage pipe, turning the water a garish purple. This canal, which empties into the Yellow River, once teemed with fish and turtles, he says. Now its water is too toxic to use even for irrigation; two of Shen's goats died within hours of drinking from the canal.

The deadly pollution comes from the phalanx of chemical and pharmaceutical factories above Shen's fields, in Shizuishan, now considered one of the most polluted cities in the world. A robust man with a salt-and-pepper crew cut, Shen has repeatedly petitioned the environmental bureau to stop the unregulated dumping. The local official in charge of enforcement responded by deeming Shen's property "uninhabitable." *(Continued on page 62)*

Adapted from "Bitter Waters" by Brook Larmer: National Geographic Magazine, May 2008.

"When the Yellow River is at peace, China is at peace." So says the message on the Sanmenxia Dam, completed in 1960 to help stop chronic flooding. Instead, it slowed the river's current, increasing siltation—and flooding. The only way to fix the dam, says one of its original engineers, is to blow it up.

(Continued from page 59) Declaring that nothing else could be done, the official then left for a new job promoting the very industrial park he was supposed to be policing. "We are slowly poisoning ourselves," says Shen, shaking with anger. "How can they let this happen to our Mother River?"

Few waterways capture the soul of a nation more deeply than the Yellow, or the Huang, as it's known in China. It is to China what the Nile is to Egypt: the cradle of civilization, a symbol of enduring glory, a force of nature both feared and revered. From its mystical source in the 14,000-foot Tibetan highlands, the river sweeps across the northern plains where China's original inhabitants first learned to till and irrigate, to make porcelain and gunpowder, to build and bury imperial dynasties. But today, what the Chinese call the Mother River is dying. Stained with pollution, tainted with sewage, crowded with ill-conceived dams,

it dwindles at its mouth to a lifeless trickle. There were many days during the 1990s that the river failed to reach the sea at all.

The demise of the legendary river is a tragedy whose consequences extend far beyond the more than 150 million people it sustains. The Yellow's plight also illuminates the dark side of China's economic miracle, an environmental crisis that has led to a shortage of the one resource no nation can live without: water.

Water has always been precious in China, a country with roughly the same amount of water as the United States but nearly five times the population. The shortage is especially acute in the arid north, where nearly half of China's population lives on only 15 percent of its water. These accidents of history and geography made China vulnerable; a series of man-made shocks are now pushing it over the edge. Global warming is accelerating the retreat of the glaciers that feed China's major rivers even as it hastens the advance of deserts that now swallow up a million acres of grassland each year.

Largest city on the Yellow River, Lanzhou is also among China's most polluted urban areas. During the past 2,500 years the waterway has overflowed and changed course more than 1,500 times, earning it the epithet "China's Sorrow."

Nothing, however, has precipitated the water crisis more than three decades of breakneck industrial growth. China's economic boom has, in a ruthless symmetry, fueled an equal and opposite environmental collapse. In its race to become the world's next superpower, China is not only draining its rivers and aquifers with abandon; it is also polluting what's left so irreversibly that the World Bank warns of "catastrophic consequences for future generations."

If that sounds like hyperbole, consider what is happening already in the Yellow River Basin. The spread of deserts is creating a dust bowl that may dwarf that of the American West in the 1930s, driving down grain production and pushing millions of "environmental refugees" off the land. The poisonous toxins choking the waterways—50 percent of the Yellow River is considered biologically dead—have led to a spike in cases of cancer, birth defects, and waterborne disease along their banks. Pollution-related protests have jumped—there were 51,000 across China in 2005 alone—and could metastasize into social unrest. Any one of these symptoms, if unchecked, could hinder China's growth and reverberate across world markets. Taken together, the long-term impact could be even more devastating. As Premier Wen Jiabao has put it, the shortage of clean water threatens "the survival of the Chinese nation."

The Yellow River's epic journey across northern China is a prism through which to see the country's unfolding water crisis. From the Tibetan nomads leaving their ancestral lands near the river's source to the "cancer villages" languishing in silence near the delta, the Mother River puts a human face on the costs of environmental destruction. But it also shows how this emergency is shocking the government—and a small cadre of environmental activists—into action. The fate of the Yellow River still hangs in the balance. *(Continued on page 66)*

PARCHED AND POLLUTED

Roughly half of China's population lives in the north, where demand for water far exceeds the natural supply. Much of the water that remains is badly tainted: The Yellow's drainage basin feeds a 3,400-mile-long river, 50 percent of whose water is deemed undrinkable.

Yellow River drainage basin (detailed at right)

Beijing

CHINA

Planned

Yangtze

THREE GORGES DAM

0 mi 250
0 km 250

•••• South-to-North Water Transfer Project (under construction)

▨ Water-stressed population, 2000

Source of the Yellow River

MONG

Shiz
Shiz (Da
Yinc
Qingtongx

Qinghai Lake

Xining

Lanzhou

QINGHAI

GANSU

Madoi

Yellow (Huang)

PLATEAU OF TIBET

POLLU

Good f
drinkin

0 mi 200
0 km 200

M. BRODY DITTE MORE AND MARGUERITE B. HUNSIKER, NG STA FF.

SOURCES: YELLOW RIVER CONSERVANCY COMMISSION; P. A. GREEN, C. J. VÖRÖSMARTY, AND COLLEAGUES , WATER SYSTEMS ANALYSIS GROUP, UNIVERSITY OF NEW HAMPSHIRE; GLOBAL LANDCOVER 2000

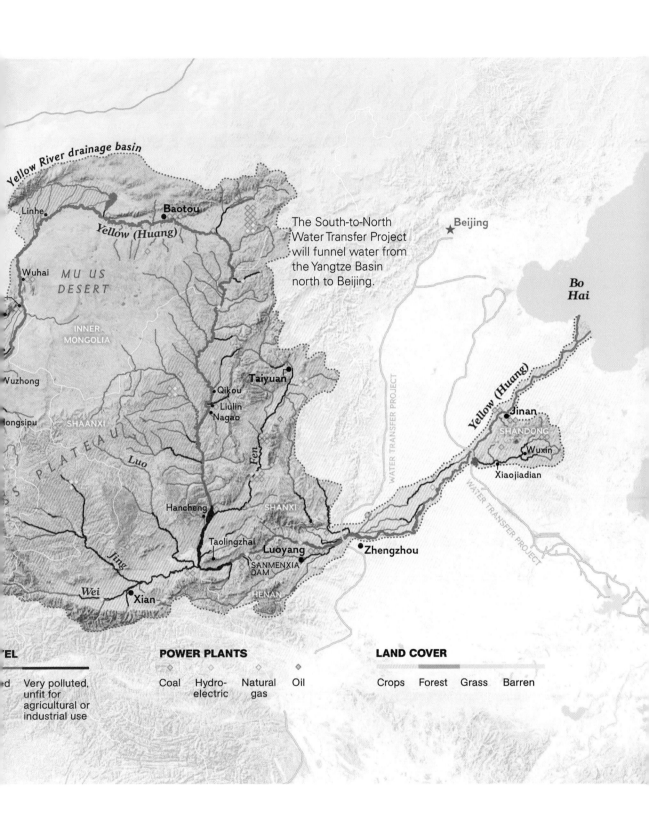

The South-to-North Water Transfer Project will funnel water from the Yangtze Basin north to Beijing.

Yellow River drainage basin

Linhe

Baotou

Yellow (Huang)

Wuhai

MU US DESERT

Wuzhong

INNER MONGOLIA

Beijing

Bo Hai

Hongsipu

SHAANXI

Qikou

Liulin
Nagao

Taiyuan

Yellow (Huang)

Jinan

SHANDONG

Wuxin

Xiaojiadian

LOESS PLATEAU

Luo

Fen

WATER TRANSFER PROJECT

WATER TRANSFER PROJECT

Hancheng

SHANXI

Jing

Taolingzhai

Luoyang

Zhengzhou

SANMENXIA DAM

Wei

Xian

HENAN

EL

Very polluted, unfit for agricultural or industrial use

POWER PLANTS

Coal   Hydro-electric   Natural gas   Oil

LAND COVER

Crops   Forest   Grass   Barren

(Continued from page 63) Sitting on a ridge nearly three miles above sea level, a rosy-cheeked Tibetan herder with two gold teeth looks out over the highlands her family has roamed for generations. It is a scene of stark beauty: rolling hills blanketed by sprouts of summer grass; herds of yaks and sheep grazing on distant slopes; and in the foreground a clear, shallow stream that is the beginning of the Yellow River. "This is sacred land," says the woman, a 39-year-old mother of four named Erla Zhuoma, recalling how her family of nomads would rotate through here to graze their 600 sheep and 150 yaks. No longer, she says, shaking her head in dismay. "The drought has changed everything."

The first signs of trouble emerged several years ago, when the region's lakes and rivers began drying up and grasslands started withering away, turning the search for her animals' food and water into marathon expeditions. Chinese scientists say the drought is a symptom of global warming and overgrazing. But Zhuoma blames the misfortune on outsiders—members of the ethnic Han Chinese majority—who angered the gods by mining for gold in a holy mountain nearby and fishing in the sacred lakes at the Yellow River's source. How else could she comprehend the death by starvation of more than half of her animals? Fearing further losses, Zhuoma and her husband accepted a government offer to sell off the rest in exchange for a thousand-dollar annual stipend and a concrete-block house in a resettlement camp near the town of Madoi. The herders are now the herded, nomads with nowhere to go.

China's water crisis begins on the roof of the world, where the country's three renowned rivers (the Yellow, the Yangtze, and the Mekong) originate. The glaciers and vast

> The proliferation of factories farms, and cities—all products of China's spectacular economic boom—is sucking the Yellow River dry.

underground springs of the Qinghai-Tibet plateau—known as China's "water tower"—supply nearly 50 percent of the Yellow River's volume. But a hotter, drier climate is sending the delicate ecosystem into shock. Average temperatures in the region are increasing, according to the Chinese weather bureau, and could rise as much as three to five degrees Celsius by the end of the century. Already, more than 3,000 of the 4,077 lakes in Qinghai Province's Madoi County have disappeared, and the dunes of the high desert lap menacingly at those that remain. The glaciers, meanwhile, are shrinking at a rate of 7 percent a year. Melting ice may add water to the river in the short term, but scientists say the long-term consequences could be fatal to the Yellow.

To save its great rivers, Beijing is performing a sort of technological rain dance, with the most ambitious cloud-seeding program in the world. During summer months, artillery and planes bombard the clouds above the Yellow River's source area with silver iodide crystals, around which moisture can collect and become heavy enough to fall as rain. In Madoi, where the thunderous explosions keep Zhuoma's family awake at night, the meteorologists staffing the weather station say the "big gun" project is increasing rainfall and helping replenish glaciers near the Yellow River's source. Local Tibetans, however, believe the rockets, by angering the gods once more, are perpetuating the drought.

Like thousands of resettled Tibetan refugees across Qinghai, Zhuoma mourns the end of an ancient way of life. The family's wealth, once measured by the size of its herds, has dwindled to the few adornments she wears: three silver rings, a stone necklace, and *(Continued on page 68)*

Low water exposes sandbars near the town of Qikou in Shanxi Province, and along many other stretches. With so much water diverted for irrigation and industry, the Yellow River ran dry before reaching its terminus—the Bo Hai sea—in all but one year during the 1990s.

To move water from southern China to water-starved parts of the north, a massive 62-billion-dollar network of concrete conduits is now under construction—including a pipe (above) that runs beneath the Yellow River. Water delivery to Beijing begins this year.

(Continued from page 66) her two gold teeth. Zhuoma has no job, and her husband, who rents a tractor to make local deliveries, earns three dollars on a good day. Not long ago the family ate meat every day; now they get by on noodles and fried dough. "We have no choice but to adjust," she says. "What else can we do?" From her concrete home, Zhuoma can still see the silvery beginnings of the Yellow River, but her relationship to the water and the land—to her heritage—has been lost forever.

"What are you doing?" the security guard demands. "Nothing," replies the stocky woman lurking outside the gates of the paper mill, tucking her secret weapon—a handheld global positioning device—under her sweater. The guard eyes her for a minute, and the woman, a 51-year-old laid-off factory worker named Jiang Lin, holds her breath. When he turns away, she pulls out the GPS and quickly locks in the paper mill's coordinates.

As an employee of Green Camel Bell, an environmental group in the western city of Lanzhou, Jiang is following up on a tip that the mill is dumping untreated chemical waste into a tributary of the Yellow River. There are hundreds of such factories around Lanzhou, a former Silk Road trading post that has morphed into a petrochemical hub. In 2006 three industrial spills here made the Yellow River run red. Another turned it white. This one is tainting the tributary a toxic shade of maroon. When Jiang gets back to the office, the GPS data will be emailed to Beijing and uploaded onto a Web-based "pollution map" for the whole world to see.

For all of Lanzhou's pride in being the first and biggest city along the Yellow River, it is better known for its massive discharge of industrial and human waste. But even here there is a glimmer of hope: the first seedlings of environmental activism, which may be the only chance for the river's salvation. In the mid-1990s a mere handful of environmental groups existed in China. Today there are several thousand, including Green Camel Bell. Jiang Lin's 25-year-old son, Zhao Zhong, founded the group in 2004 to help clean up the city and protect the Yellow River. With only five paid staff, Green Camel Bell is a shoestring operation kept afloat by grants from an American NGO, Pacific Environment. The name they chose, after the reassuring bells worn by camels in Silk Road caravans, is meant to be "a sign of life," says Jiang. "The bell is supposed to give hope to everyone who hears it."

At long last Beijing appears willing to listen. After three decades blindly pursuing growth, the government is starting to grapple with the environmental costs. The impact is not simply monetary, though the World Bank calculates that environmental damage robs China of 5.8 percent of its GDP each year. It is also social: Irate citizens last year flooded the government with hundreds of thousands of official environmental complaints. Whether to save the environment or stave off social unrest, Beijing has adopted ambitious goals, aiming for a 30 percent reduction in water consumption and a 10 percent decrease in pollution discharges by 2010.

Yet despite the good intentions, the crisis is only getting worse, reflecting Beijing's loss of control over the country's growth-hungry provinces. Leading environmental lawyer Wang Canfa estimates that "only 10 percent of environmental laws are enforced." Unable to count on its own bureaucracy, Beijing has warily embraced the media and grassroots activists to help pressure local industry. But pity the ecological crusader who speaks out too much. He could end up like Wu Lihong, an activist who was jailed and allegedly

Only 10 percent of environmental laws are enforced.

tortured last year for publicizing the toxic algal blooms in central China's Tai Lake.

Back in the Green Camel Bell office, Jiang stresses the group's cordial relations with local authorities. "The government has been working hard to stop factories from dumping," she says. Nevertheless, along her office wall stand plastic bottles filled with water discharged by factories and ranging in color from yellow to magenta— all unanalyzed for lack of funds. Even with its modest resources, Green Camel Bell has mobilized volunteers to help survey the ecology of the 24-mile section of the Yellow River that flows through Lanzhou. Their most important, and stealthiest, work is publicly exposing the most egregious polluters. It's enough to give a laid-off worker a sense of power and purpose. "I feel like a detective," says Jiang, laughing about her narrow escape at the paper mill. "But ordinary people like me have to get involved. Pollution is a problem that affects us all."

Two hundred miles northeast of Lanzhou, the Yellow River carves a path through the desolate expanse of Ningxia, revealing a problem with even more devastating long-term consequences than pollution: water scarcity. China starts at a disadvantage, supporting 20 percent of the world's population with just 7 percent of its fresh water. But it is far worse here in Ningxia, a bone-dry region enduring its worst drought in recorded history. For millennia the Yellow River was Ningxia's salvation; today the waterway is wasting away. Near the city of Yinchuan, the river's once mighty current is reduced to a narrow channel. Locals blame the river's depletion on the lack of rain. But the biggest culprit is the extravagant misuse of water by rapidly expanding farms, factories, and cities.

Perhaps every revolution, even a capitalist one, eats its children. But the pace at which China is squandering its most precious resource is staggering. *(Continued on page 72)*

On an arid expanse of the Loess Plateau, farmer Ren Guibao trudges uphill with cornstalks he'll burn as fuel. Rare rains wash the loose yellow soil down gullies to the river, tinting the water its namesake color. Sediments choke some stretches of the Yellow, hindering its ability to flush out pollutants.

(*Continued from page 69*) Judicious releases of reservoir water have averted the embarrassment of recent years, when the Yellow River ran completely dry. But the river's outflow remains just 10 percent of the level 40 years ago. Where has all the water gone? Agriculture siphons off more than 65 percent, half of which is lost in leaky pipes and ditches. Heavy industry and burgeoning cities swallow the rest. Water in China, free until 1985, is still so heavily subsidized that conservation and efficiency are largely alien concepts. And the siege of the Yellow River isn't about to stop: In 2007 the government approved 52 billion dollars in coal mining and chemical industries to be installed along a 500-mile stretch of the river north of Yinchuan.

Such frenzied growth may soon fall victim to the very water crisis it has helped create. Of the some 660 cities in China, more than 400 lack sufficient water, with more than a hundred of these suffering severe shortages. (Beijing is chronically short of water too, but it will be spared during the Olympics, thanks to engineering feats that divert water from the Yellow River.) In a society increasingly divided between urban and rural, rich and poor, it is China's vast countryside—and its 738 million peasants—that bears the brunt of the water shortage.

The lack of water is already hindering China's grain production, fueling concerns about future shocks to global grain markets, where even modest price hikes can have a disastrous effect on the poor. Wang Shucheng, China's former minister of water resources, put the situation dramatically: "To fight for every drop of water or die, that is the challenge facing China."

For Sun Baocheng, a sunbaked 37-year-old farmer from the central Ningxia village

To save its great rivers, Beijing is performing a sort of technological rain dance, with the most ambitious cloud-seeding program in the world.

of Yanghe, this challenge is not merely rhetorical excess. Two years ago, after their wells and rain buckets went dry from drought, all 36 families in Yanghe abandoned their village to the encroaching desert. They came to a valley called Hongsipu, where more than 400,000 environmental refugees have settled for one reason: It has water, delivered by a Kuwaiti-funded aqueduct that snakes across the scrub desert from the Yellow River, 20 miles to the north. The Yanghe villagers have settled in a row of single-room brick houses near the concrete aqueduct, tending plots of land given by the Chinese government (along with about $25 a person) as part of a program to alleviate poverty and desertification.

Even though Sun is barely able to coax a few stalks of corn out of the sandy soil, he is inspired by the flourishing crops—and growing wealth—of more established refugees. "If we hadn't left our old village and come here," he says, "we wouldn't have survived." The Mother River, once again, is giving life. But with all the pressures on its dwindling water, one wonders: What will creating another oasis in the desert do to the river's own chances of survival?

Mao Zedong's mantra—"Sacrifice one family, save 10,000 families"—is still seared into Wang Yangxi's memory. Like the Chinese emperors before him, Chairman Mao was obsessed with taming the Yellow River, the life-giving force whose changes of course also unleashed devastating floods, earning it the enduring sobriquet "China's Sorrow." When, in 1957, construction began on the massive dam at Sanmenxia, on the river's middle section, 400,000 people—including Wang—lost their homes. Mao's slogan convinced them it was a noble sacrifice.

Shooting for rain, meteorologist Huang Binming (right) and his crew launch canisters of silver iodide into the sky near Madoi. This cloud-seeding program has increased rainfall close to the river's source, says Binming, though some Tibetans complain these noisy fusillades anger the gods. "People in our city don't think that way," says Binming. "They are very cooperative."

"We were proud to help the national cause," says Wang, now 83. "We've had nothing but misery ever since."

The idea of conquest has driven China's approach to nature ever since Yu the Great, first ruler of the Xia dynasty, allegedly declared some 4,000 years ago: "Whoever controls the Yellow River controls China." Mao took this, like much else, to extremes. His biggest monument to man's power over nature—the 350-foot-tall Sanmenxia Dam—is a case study in the danger of unintended consequences. The dam has tamed the lower third of the Yellow River by turning it into what one commentator has called "the country's biggest irrigation ditch." But the impact upriver has been disastrous, due to a stunning lack of foresight. Engineers failed to account for the colossal amount of yellowish silt (more than three times the sediment discharge of the Mississippi) that gives the river its name. By mismanaging the silt, Sanmenxia has caused as many floods as it has prevented, ruined as many lives as it has saved, *(Continued on page 78)*

No longer willing to drink putrid river water, residents of Nagao use long plastic tubes to tap a single well. China's groundwater usage has almost doubled since 1970. Today two-thirds of the nation's total water consumption comes from aquifers, and the water table keeps falling.

The drive to consume clouds the road ahead in Liulin, where a coal-fired power plant helps fuel an energy-hungry China. Environmentalists say that greenhouse gases are contributing to the death of the Yellow River, but economists warn that a slowdown in China could lead to political instability—even chaos.

(*Continued from page 73*) and compelled the construction of another huge dam simply to correct its mistakes. One of Sanmenxia's original engineers even recommends blowing up the whole thing.

Wang would be the first to volunteer for such a mission. Husking cotton on his doorstep in Taolingzhai village, about 30 miles west of Sanmenxia, the bristle-haired former schoolteacher recalls a life whose every tragic twist has been shaped by the dam. After Wang and his family were evicted from this fertile land during the dam's construction, they were banished to a desert region 500 miles away. Nearly a third of the refugees died of starvation during Mao's Great Leap Forward, he says. Eventually, half of the survivors straggled home. Wang now farms land near the junction of the Wei and Yellow Rivers. But even here, he is not safe. When heavy rains fall, the Sanmenxia reservoir backs up, pushing polluted water over the banks. Three floods in five years have destroyed his cotton crops and poisoned the village's drinking supply. "All of our young people have left," says Wang. "There's no future here."

Unlike Mao's little red book, the Sanmenxia Dam is hardly a relic of the past. China now boasts nearly half of the world's 50,000 large dams—three times more than the United States—and construction continues. A cascade of 20 major dams already interrupt the Yellow River, and another 18 are scheduled to be built by 2030. Grassroots resistance to dams has emerged, most famously over the forced resettlement of more than a million people by the Yangtze River's Three Gorges Dam, but to little effect. Ma Jun, a prominent environmentalist, says dams on the Yellow River are especially harmful, since they exacerbate the twin threats of pollution and scarcity. The reduced water flow destroys the river's ability to flush out heavy pollutants, even as standing reservoirs allow a badly overused

After three decades blindly pursuing growth, China's government is starting to grapple with the environmental costs.

river to be drained even further. "Why cannot human beings give up their ruthless ambition of harnessing and controlling nature," Ma asks, "and choose instead to live in harmony with it?"

The simple answer: Beijing is still addicted to growth. The economic boom has lifted hundreds of millions of Chinese out of poverty, and the Communist Party's legitimacy, perhaps even its survival, depends on continued expansion. China's leaders pay lip service to conservation and efficiency as a solution to the north's chronic water shortage. But rather than raise the price of water to true market levels—a move that would surely alienate both the masses and big industry—they have opted instead for another pharaonic feat of engineering: the South-to-North Water Transfer Project. The 62-billion-dollar canal system, which is designed to relieve pressure on the Yellow River, will siphon some 12 trillion gallons of water a year from the Yangtze Basin and send it 700 miles north, passing beneath the Yellow in two places. It's no surprise, given the Olympian scale of the project, that it—like Sanmenxia—originated as one of Mao's pipe dreams.

Even as other parts of China careened through droughts and floods in past decades, the village of Xiaojiadian enjoyed a steady supply of fresh water by virtue of its location on a tributary of the Yellow River, less than 200 miles from where it spills into the sea. But the waters, once a source of life, have turned deadly. Nobody here likes to talk about the plague that has struck the village, but the scar running down the chest of a gaunt farmer named Xiao Sizhu has its own eloquence. It shows precisely where doctors tried to remove the cancerous tumor gnawing at his esophagus. In between bites of sodden bread—one of the only foods he can digest—Xiao, 55, whispers about the old days, when his family felt lucky to live in this well-watered corner of the river basin, in eastern Shandong Province.

Keeping your balance while moving ahead—that's the trick for children playing inside plastic bubbles on a lake in Shizuishan. It's also the challenge for China, where balance is a Confucian value—"harmony" is a national watchword—but where an environmental crisis poses a global question: Can we keep on growing without destroying ourselves?

Over the past two decades, however, a parade of tanneries, paper mills, and factories arrived upstream, dumping waste directly into the river. Xiao used to swim and fish in the eddy next to the village well. Now, he says, "I never go close to the water because it smells awful and has foam on top."

Another place he avoids is the grove of poplar trees outside the village, with its burial mounds stretching to the river's edge. In the past five years more than 70 people in this hamlet of 1,300 have died of stomach or esophageal cancer. More than a thousand others in 16 neighboring villages have also succumbed. Yu Baofa, a leading Shandong oncologist who has studied the villages of Dongping County,

calls it "the cancer capital of the world." He says the incidence of esophageal cancer in the area is 25 times higher than the national average.

The more than four billion tons of wastewater dumped annually into the Yellow River, accounting for a full 10 percent of the river's volume, has pushed into extinction a third of the river's native fish species and made long stretches unfit even for irrigation. Now comes the human toll. In a 2007 report China's Ministry of Health blamed air and water pollution for an alarming rise in cancer rates across China since 2005—19 percent in urban areas and 23 percent in the countryside. Nearly two-thirds of China's rural population, more than

500 million people, use water contaminated by human or industrial waste. It's little wonder that gastrointestinal cancer is now the number one killer in the countryside.

The ubiquity of pollution-related disease is cold comfort to the villagers in Xiaojiadian, who live in fear and shame. The fear is understandable: 16 more cases of cancer were diagnosed in the village last year. The shame, however, has deeper roots. Even though officials told villagers the epidemic likely stems from the drinking well by the poisoned river, many locals believe cancer comes from an imbalance of chi, or life force, which is said to occur more frequently in those with quick tempers or bad characters.

Like most victims, Xiao suffered in silence in his house for nearly a year, hiding his symptoms even from the local doctor. Medical bills have since wiped out his savings, and the tumor has reduced his voice to a whisper. Even so, Xiao is one of the few willing to speak out. "If we don't talk, nothing gets done," he rasps, spitting up phlegm into a plastic cup. The government recently built a new well 11 miles away and sent in teams of doctors. But Xiao says officials might not have paid attention to Xiaojiadian had a villager not tipped off a reporter at a Chinese television station two years before. Now Xiao only has one regret: that he didn't speak out earlier. "It might have saved me," he says.

A few months pass, and a fresh earthen mound appears in the grove of poplar trees by the river. The grave has no tombstone, just some bamboo sticks and a few aluminum cookie wrappers rustling in the breeze. Xiao has come to the place he long avoided, joining friends and neighbors who were stalked by the same waterborne assassin. Is it a cruel irony or just the natural order that their final resting place overlooks the very river that likely killed them?

It is too late to save Xiao Sizhu, but there remains a flicker of hope that the Yellow River can be rescued. China's leaders, aware of the peril their country faces, now vow "to build an ecological civilization," setting aside almost 200 billion dollars a year for the environment. But the future depends equally on ordinary citizens such as activists Zhao Zhong and his mother, the intrepid Jiang Lin. Remember that Lanzhou paper mill Jiang locked in with her GPS? Not long after the information went up on the Internet, the government shut down the mill, along with 30 other factories dumping poison into tributaries of the Yellow River.

"Maybe the impact of one single person is small," says Zhao. "But when it is combined with others, the power can be huge."

Discussion Questions

- Why are water quality, quantity and distribution so important for China? How has China's use of its major rivers helped and sometimes hurt its economic development?

- Describe some of the specific ecological and social results of the widespread pollution that is dumped into the Yellow River.

- What is cloud seeding? How does it work? Do you personally think that it is a good strategy for addressing the problem of a decrease in precipitation levels?

- The article mentions that the World Bank calculates that environmental damage robs China of 5.8% of its GDP each year. Explain how pollution can have a negative effect on the economy and how conservation practices that cost money initially can sometimes result in a more robust economy over the long term.

- What does it mean to have the government subsidize water? What are the positive and negative effects of this subsidy? Do you think that it is a good idea for governments to subsidize the use of scarce natural resources?

Map Literacy

Use the map contained in this article to explain the following:

- Based on the map, what percent of the Yellow River and its tributaries do you estimate to be suitable for drinking? What percent is so polluted that it cannot even be used for industrial purposes?

- How does the area of the country whose population is stressed as a result of water availability relate to the level of pollution? Does the water tend to be more polluted in areas of high water stress or low water stress? Explain why you think this is.

PARTING THE WATERS

Parting the Waters describes the implications of water policy on international relations in the Middle East. It illustrates the importance of the Jordan River in not only the economic and ecological roles that it plays in the regional ecosystem, but also the essential social and political roles that it plays in the Israeli/Palestinian conflict. Freshwater sources have played an important part in international affairs throughout history in sometimes negative and sometimes positive ways. This article illustrates a good case study of how a river can be a source of regional conflict, but also a basis for partnership in the interest of conserving shared resources.

When reading this article, you should focus on:

- Why the Jordan River is such an important resource for regional countries,
- Why the river is so polluted and has lost so much of its historic flow,
- Why so many military conflicts have been fought over it and how inequities in the amount used by individual countries has potentially played a role,
- How the necessity of freshwater stewardship can sometimes promote dialogue and cooperation between opposing sides, and
- What role these inequities and cooperative studies may play in future peace discussions.

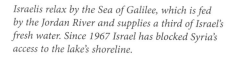

Israelis relax by the Sea of Galilee, which is fed by the Jordan River and supplies a third of Israel's fresh water. Since 1967 Israel has blocked Syria's access to the lake's shoreline.

PARTING THE WATERS

By Don Belt

Photographs by Paolo Pellegrin

Israeli border police (foreground) stand guard on the western
bank of the lower Jordan, where pilgrims visit the river where
Jesus is believed to have been baptized. On the Jordanian side,
a tourist center and several churches commemorate the site.

A SOURCE OF CONFLICT BETWEEN ISRAEL AND ITS NEIGHBORS FOR DECADES, THE JORDAN RIVER IS NOW DEPLETED BY DROUGHT, POLLUTION, AND OVERUSE. COULD THE FIGHT TO SAVE IT FORGE A PATH TOWARD PEACE?

Watch your step, my friend, and whatever you do, don't step on a bleeping mine.

For a biblical stream whose name evokes divine tranquillity, the Jordan River is nobody's idea of peace on Earth. From its rowdy headwaters near the war-scarred slopes of Mount Hermon to the foamy, coffee-colored sludge at the Dead Sea some 200 miles downstream, the Jordan is fighting for survival in a tough neighborhood—the kind of place where nations might spike the riverbank with land mines, or go to war over a sandbar. Water has always been precious in this arid region, but a six-year drought and expanding population conspire to make it a fresh source of conflict among the Israelis, Palestinians, and Jordanians vying for the river's life-giving supply.

All of which makes the scene one morning last July all the more remarkable. Accompanied by military escort, three scientists—an Israeli, a Palestinian, and a Jordanian—are standing knee-deep in the Jordan River. They are nearly 40 miles south of the Sea of Galilee, under the precarious ruins of a bridge that was bombed during the Six Day War of June 1967. The scientists are surveying the river for Friends of the Earth Middle East (FOEME), a regional NGO dedicated to building peace through environmental stewardship. It's a scorching hot day in a former war zone, but if these men are concerned about the danger of heat stroke, getting clonked by a chunk of falling concrete, or stepping on a mine washed downstream by a flood, they're hiding it well.

"Hey, Samer," says Sarig Gafny, an Israeli ecologist in a floppy, green hat, "check this little fellow out." Samer Talozi, a tall, self-possessed young environmental engineer from Jordan, peers over his shoulder at the tiny invertebrate his Israeli colleague has scooped into a glass sample jar. "It lives!" he says with a laugh. "That is one tough crustacean!" A few yards away, Banan Al Sheikh, a stout, good-natured botanist from the West Bank, is absentmindedly wading upstream while focusing his

Adapted from "Parting the Waters" by Don Belt: National Geographic Magazine, April 2010.

The Romans built pools on the Baniyas River, one of the Jordan's headwaters in the Golan Heights. Syria had sought to divert the Baniyas before Israel captured the Golan.

camera on a flowering tree amid the tall reeds and other riparian species along the riverbank. "Watch your step, my friend," Gafny calls out after him, "and whatever you do, don't step on a bleeping mine."

Besides lethal munitions, this stretch of the Jordan River—perhaps 25 feet wide and a few feet deep—is so polluted that any sign of aquatic life is worth celebrating. Part of the reason is water scarcity: In the past five decades the Jordan has lost more than 90 percent of its normal flow. Upstream, at the Sea of Galilee, the river's fresh waters are diverted via Israel's National Water Carrier to the cities and farms of Israel, while dams built by Jordan and Syria claim a share of the river's tributaries, mostly for agriculture. So today the lower Jordan is practically devoid of clean water, bearing instead a toxic brew of saline

water and liquid waste that ranges from raw sewage to agricultural runoff, fed into the river's vein like some murky infusion of tainted blood.

The fight over the Jordan illustrates the potential for conflict over water that exists throughout the world. We live on a planet where neighbors have been clubbing each other over rivers for thousands of years. (The word "rival," from the Latin rivalis, originally described competitors for a river or stream.) Worldwide, a long list of watersheds brims with potential clashes: between India and Pakistan over the Indus; Ethiopia and Egypt over the Nile; Turkey and Syria over the Euphrates; Botswana and Namibia over the Okavango. Yet according to researchers at Oregon State University, of the 37 actual military conflicts over *(Continued on page 89)*

Bearing the burdens of a dry land, the Jordan south of the Sea of Galilee is mostly saline water and liquid waste, its fresh waters pumped out upstream for farming and daily use.

LIFELINE IN THE HOLY LAND

In quieter parts of the world, the 200-mile-long Jordan might be considered a minor stream. But here, coveted by rivalrous neighbors in a rain-starved region, the river has sparked more than its share of conflicts—and occasionally, cooperation.

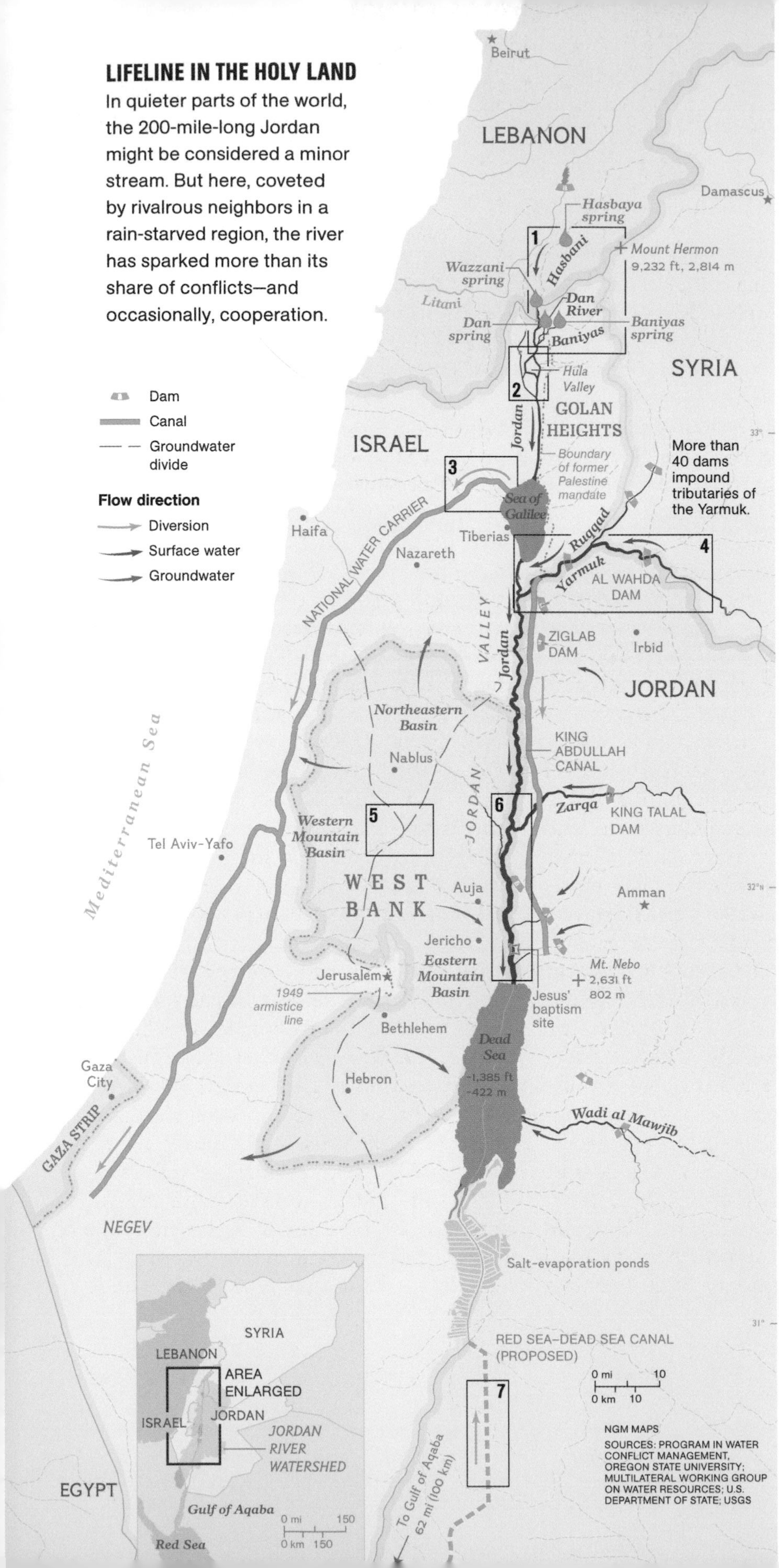

Dam
Canal
Groundwater divide
Flow direction
Diversion
Surface water
Groundwater
Beirut
LEBANON
Damascus
Hasbaya spring
Hasbani
Mount Hermon 9,232 ft, 2,814 m
Wazzani spring
Litani
Dan River
Dan spring
Baniyas
Baniyas spring
SYRIA
Hula Valley
GOLAN HEIGHTS
Jordan
Boundary of former Palestine mandate
ISRAEL
More than 40 dams impound tributaries of the Yarmuk.
Sea of Galilee
Haifa
Tiberias
Ruqqad
NATIONAL WATER CARRIER
Nazareth
Yarmuk
AL WAHDA DAM
VALLEY
ZIGLAB DAM
Irbid
JORDAN
Northeastern Basin
Mediterranean Sea
KING ABDULLAH CANAL
Nablus
Zarqa
KING TALAL DAM
Western Mountain Basin
Tel Aviv-Yafo
WEST BANK
Auja
Amman
Jericho
Eastern Mountain Basin
Mt. Nebo 2,631 ft 802 m
Jerusalem
1949 armistice line
Jesus' baptism site
Bethlehem
Dead Sea
-1,385 ft -422 m
Gaza City
Hebron
GAZA STRIP
Wadi al Mawjib
NEGEV
Salt-evaporation ponds
SYRIA
LEBANON
AREA ENLARGED
ISRAEL
JORDAN
JORDAN RIVER WATERSHED
EGYPT
Gulf of Aqaba
Red Sea
0 mi 150
0 km 150
RED SEA–DEAD SEA CANAL (PROPOSED)
0 mi 10
0 km 10
To Gulf of Aqaba 62 mi (100 km)
NGM MAPS
SOURCES: PROGRAM IN WATER CONFLICT MANAGEMENT, OREGON STATE UNIVERSITY; MULTILATERAL WORKING GROUP ON WATER RESOURCES; U.S. DEPARTMENT OF STATE; USGS

1 HEADWATERS
From springs around Mount Hermon, three rivers converge in Israel to form the Jordan. After 1948 Israel treated any upstream diversion by Syria or Lebanon as a hostile act.

2 HULA VALLEY
To boost agriculture in the 1950s, Israel drained swamps bordering the Syrian Golan Heights. Skirmishes there continued until 1967, when Israel captured the Golan.

3 NATIONAL WATER CARRIER
Completed by Israel in 1964 despite fierce Arab opposition, the canal was built to move water from the Sea of Galilee to Tel Aviv and farms in the Negev desert.

4 YARMUK RIVER
Largest tributary of the Jordan, the Yarmuk is tapped by Syria, Jordan, and Israel. Secret talks between Israel and Jordan over its water foreshadowed a peace agreement in 1994.

5 GROUNDWATER
Israel's occupation of the West Bank after 1967 gave it control of the area's three major aquifers, or basins; negotiations over groundwater began during the Oslo peace talks in the 1990s.

6 LOWER JORDAN
Partly an international border and used as a waste canal, the lower Jordan is flanked by military zones and minefields and is so polluted that it hardly supports life.

7 RED SEA–DEAD SEA CANAL
Barely replenished by the Jordan, the Dead Sea has fallen to alarmingly low levels. One controversial solution: a canal connecting it to the Red Sea.

(Continued from page 86) water since 1950, 32 took place in the Middle East; 30 of them involved Israel and its Arab neighbors. Of those, practically all were over the Jordan River and its tributaries, which supply millions of people with water for drinking, bathing, and farming.

Armed confrontations over the Jordan date to the founding of Israel in 1948 and the recognition that sources of the country's needed water supply lay outside its borders. Its survival depended on the Jordan River, with its headwaters in Syria and Lebanon, its waters stored in the Sea of Galilee, and the tributaries that flow into it from neighboring countries.

Israel's neighbors face a similar situation. Their survival is no less at stake—which makes the line between war and peace here very fine indeed. In the 1960s Israeli air strikes after Syria attempted to divert the Baniyas River (one of the Jordan's headwaters in the Golan Heights), together with Arab attacks on Israel's National Water Carrier project, lit fuses for the Six Day War. Israel and Jordan nearly came to blows over a sandbar in the Yarmuk River in 1979. And in 2002 Israel threatened to shell agricultural pumping stations on the Hasbani, another of the headwaters in southern Lebanon.

Yet fights over water have also led to dialogue. "There are few major sources of water that don't cross one or more political boundaries," says Gidon Bromberg, the Israeli co-director of Friends of the Earth Middle East. "That creates a natural interdependence between countries." Sharing resources can actually be a path to peace, Bromberg says, because it forces people to work together.

Israel and its neighbors face a similar situation: Their survival is at stake—which makes the line between war and peace very fine indeed.

In the 1970s, for example, Jordan and Israel agreed on how to divvy up water even when the countries were officially at war. And cooperation between Israelis and Palestinians over water has continued even as other tracks of the peace process hit a wall.

"It seems counterintuitive, but water is just too important to go to war over," says Chuck Lawson, a former U.S. official who worked on Israeli-Palestinian water issues in the 1990s. "Regardless of the political situation, people need water, and that's a huge incentive to work things out."

One day last April, Bromberg led me to the natural spring that provides water to Auja, a Palestinian village of 4,500 people that climbs the barren hills a few miles west of the Jordan River near Jericho. Fed by winter rains, the spring was flowing from a small, boulder-strewn oasis, and we trekked along the narrow concrete trough that transports water to the village, several miles away. "Auja is totally dependent on this water for agriculture," Bromberg said. "As soon as this spring dries up, there'll be no more water for farming."

Part idealist, part political operative, Bromberg was born in Israel and raised in Australia, then returned to Israel in 1988 to help build peace in the region. By challenging his own country to share water equitably, Bromberg has rattled the cages of hard-line Israeli politicians who see water as a national security issue—and as a resource to guard jealously.

Since occupying the West Bank in 1967, Israel has built a few dozen settlements in the Jordan Valley, in addition to the 120 or so elsewhere in the West Bank. (Continued on page 95)

After six years of drought, measuring sticks are useless at the Ziglab Dam in Jordan, built to catch water flowing west into the Jordan River for irrigation. Its reservoir has shrunk to a fifth of capacity and hasn't filled since 2003, forcing Jordan to ration water.

Swaddled against the sun, workers from Thailand harvest bananas on an Israeli kibbutz in the Jordan Valley. Though lucrative, the tropical import needs at least eight times as much water as tomatoes. "In a desert, that's crazy," says Gidon Bromberg of Friends of the Earth Middle East.

A source of friction between Israelis and Palestinians, water is emblematic of their unequal relationship.
At a water park in Tiberias, Israelis bask in its relative abundance (below), while Palestinians, restricted to shallow wells by
Israel's occupation, buy West Bank groundwater from Israel with European Union aid.

(Continued from page 89) The settlers' water is provided by Mekorot, Israel's national water authority, which has drilled 42 deep wells in the West Bank, mainly to supply Israeli cities. (According to a 2009 World Bank report, Israelis use four times as much water per capita as Palestinians, much of it for agriculture. Israel disputes this, arguing that its citizens use only twice as much water and are better at conserving it.) In any case, Israel's West Bank settlements get enough water to fill their swimming pools, water their lawns, and irrigate miles of fields and greenhouses.

In contrast, West Bank Palestinians, under Israeli military rule, have been largely prevented from digging deep wells of their own, limiting their water access to shallow wells, natural springs, and rainfall that evaporates quickly in the dry desert air. When these sources run dry in the summer, Bromberg said, Auja's Palestinians have no choice but to purchase water from Israel for about a dollar a cubic yard—in effect buying back the water that's been taken out from under them by Mekorot's pumps, which also lower the water table and affect Palestinian springs and wells.

As Bromberg and I followed the Auja spring east, we passed a complex of pumps and pipes behind a barbed-wire fence—a Mekorot well, drilled 2,000 feet deep to tap the aquifer. "Blue and white pipes," Bromberg said. "This is what water theft looks like in this part of the world."

Israel's chief water negotiator, Noah Kinnarti, disagrees. Underground water knows no borders, he says, and points out that Israelis must also purchase the water they use. "Palestinians think any rain that falls in the West Bank belongs to them," he told me at his kibbutz near the Sea of Galilee. "But in the Oslo talks, we agreed to share that water. They just can't get their act together to do it."

FOEME began confronting these tough issues in 2001, during a period of intense Palestinian-Israeli violence. But by focusing first on ways to improve water quality, the NGO mobilized support and built trust through its Good Water Neighbors program, a grassroots education initiative. It's also working to establish a Jordanian-Israeli peace park on a midstream island. Perhaps most important, it has pressured governments to live up to the water-sharing commitments embedded in the region's peace agreements, seeking to make the Jordan River a model for the kind of cooperation needed to avert future water wars.

"People all over the world associate the Jordan River with peace," says Munqeth Mehyar, FOEME's co-director in Jordan. "We're just helping it live up to its reputation!"

When I returned to Auja in early May, its spring had been reduced to a trickle, leaving the village as dry as a fistful of talcum powder. The fields around it lay empty and exhausted, while on Auja's one plot of flat ground, boys were playing soccer amid a swirling dust cloud they were kicking up, chasing an old leather ball worn to the consistency of flannel.

I stopped by the home of an elderly farmer named Muhammad Salama. "We haven't had running water in my house for five weeks," Salama said. "So now I have to buy a tank of water every day from Mekorot to supply my family and to water my sheep, goats, and horses." He also has to buy feed for his animals because there is no water to irrigate crops. To meet these costs he is selling off his livestock, and his sons have taken jobs at an Israeli settlement, tending the tomatoes, melons, and other crops irrigated from the aquifer that is off-limits to Palestinian farmers. "What can we do?" he asked, pouring me a glass of Mekorot water from a plastic bottle. "It's not fair, but we're powerless to do anything about it."

It was a clear day, and from his front window we could see across the parched, brown valley all the way to the thin line of gray-green vegetation marking the path of the Jordan River. For a moment, its water seemed within reach. "But to get there I'd have to jump an electric fence, cross a minefield, and fight the Israeli army," Salama said. "I'd have to start a water war!"

Tempers flare near Auja when a conversation between Israeli settlers and a local Palestinian turns to ownership of land—and water. Fed by a natural spring, Auja's only water channel for farming runs dry every summer.

Floating on dreams and whispers, girls from a West Bank
village cool off in the salt-laden waters of the Dead Sea. With
its main tributary, the Jordan, at less than a tenth of its former
volume, the inland sea has dropped some 70 feet since 1978.

Discussion Questions:

- Why has the Jordan River, along with its headwaters and tributaries, incited so many military conflicts in the last century? What makes it such an important social, political, economic and ecological resource?

- Do you think that having to share an important resource like a major river in an arid region more often helps to promote compromise and cooperation, or conflict and a constant source of tension? Cite a historical incident where scarce resources have helped result in constructive dialogue and one where they have led to military conflict.

- Do you think that NGOs like Friends of the Earth Middle East exemplify a model that can be used to bring conflicting groups together in other areas of the world? What other regions and natural resources have this potential?

- Why do the rights to and the distribution of water appear to be so imbalanced between the various sides? What role do you think this current inequity has on the potential for future negotiated peace?

Map Literacy

Use the map contained in this article to explain the following:

- What are the major differences in water quality and amount that flows through the Upper Jordan River near its headwaters and the Lower Jordan River near the Dead Sea? What are some of the reasons for these differences?

- What countries would have to play a role in a potential future agreement to manage and conserve the Jordan River? Which ones appear to have constructed the biggest or greatest number of projects to control or redistribute water from its original path?

REUNITING A RIVER

The Klamath River has been the lifeblood for surrounding communities for generations. Residents have depended on it for fish, agriculture and, more recently, a dependable source of electricity from the many dams that line the river valley. However, stagnation and reduced water flows, as a result of dams, pollution, changing climates, growing populations and overfishing, have led to a steady decline in the health of the Klamath. Concerns over its long-term vitality have led to talks about restoring sections of the river by removing some of the dams. While these kinds of discussions of large-scale restorations are controversial, they have also stimulated collaboration and better understanding between previously divided users.

When reading this article, you should focus on:
- What makes the Klamath River different from many other river systems,
- How droughts and competing water needs have led to past conflicts,
- What effect the Klamath Reclamation Project had on the river ecology and the relationship between fishermen and farmers,
- What effect removing some of the dams may have on salmon recruitment and water quality, and
- How restoration talks can help get people from very different backgrounds working together in the interest of freshwater stewardship.

Ron Reed hoists a chinook salmon into fellow Karuk Tribe member Johnny Garrison's waiting hands. Ishi Pishi Falls is the last place on the Klamath River, which straddles the Oregon-California line, where Karuk still use dip nets. Salmon have been sparse in recent years: The tribe harvests only about 400 fish a year here.

REUNITING
A RIVER

By Russ Rymer

Photographs by David Mclain

Copco No. 1 Dam is one of several Klamath dams in Oregon and California that together provide clean power for up to 70,000 homes. But dams block salmon runs and may degrade river water quality. Conservationists and Indian tribes want to raze four of them—an unprecedented removal project.

AFTER FIGHTING FOR YEARS OVER ITS WATER,
FARMERS, INDIANS, AND FISHERMEN ARE JOINING FORCES
TO LET THE TROUBLED KLAMATH RIVER RUN WILD AGAIN.

Silver shapes glinted up at Thomas Willson out of the river depths, shining like spilled coins through the surface rills. Before his square-nose aluminum skiff even reached the sandbar, Willson could tell it wouldn't be the worst of mornings, one of those days when he came up with nothing but a soiled net and went home empty-handed. But when he leaned over the gunwales and hauled the gill net up out of the strangely warm Klamath River water, what he found didn't please him:

a large chinook salmon that should have been the day's prize, except that its flanks were dull and pocked with whitish sores. When Willson ran his fingers under its gill scutes, the tissue floated out in a viscid pinkish soup. "Never used to see this," Willson grumbled, and with a discus thrower's shoulder spin he heaved the blighted carcass onto the riverbank. Above him a buzzard floated in the river canyon's narrow slice of California sky. It would soon get its commission.

Willson's expression fell on the sorrow side of anger. Fishing was more than a pastime for him and more than a vocation; it was a patrimony. In the annals of father-to-son

Fishing was more than a pastime for him and more than a vocation; it was a patrimony.

enterprises, the Willson family franchise surely ranks among the venerable: Thomas Willson and his ancestors have been fishing this very species in this very stretch of this very stream without interruption since Yurok Indians first made their home on the Klamath River and fed themselves on its salmon. Indian tribes have resided alongside the Klamath for more than 300 generations. In all that time, the river had never suffered the troubles of its recent years. The signs were everywhere: in the tresses of algae clinging to every twist and tie of his net; in the warmth of the mountain river water, which would reach 74°F before midmorning; in the smoke floating overhead from forest fires that no longer burned themselves out. And in the paucity and poor condition of the fish. The underlying source of the problems, Willson knew, was a resource crisis of growing magnitude in the western United States and globally: too many users for not enough water. Looking around him on this not worst of mornings, Willson had *(Continued on page 106)*

Adapted from "Reuniting a River" by Russ Rymer: National Geographic Magazine, December 2008.

In a holding pattern, spring chinook salmon congregate in the depths of a pool on the Salmon River. In early fall, they'll head farther up the Klamath tributary to spawn. Spring runs of chinook salmon in the Klamath numbered as high as 800,000 in the early 20th century but have fallen dramatically in recent years.

(Continued from page 103) the feeling there wasn't much about his little patch of Earth that wasn't out of balance. The Klamath River was in trouble, and Willson was certain where the trouble came from: upstream.

Two hundred and fifty miles upstream, at two in the morning, the alarm blared on the humidity meter on the Formica snack table in Steve Kandra's RV, and Kandra slid out of his berth and into his boots and climbed aboard the John Deere tractor awaiting him in his pitch-dark alfalfa field, a field irrigated by the same Klamath waters that Thomas Willson fishes. Kandra had mowed the alfalfa several days before; tonight he would bale it while the cut crop was safe from the parching daytime heat and before the morning dew turned everything too wet. Farming by ideal conditions meant living on the wrong side of adage: Kandra makes hay till the sun shines.

As drought years have become more problematic in the Klamath region, the competing water needs for Thomas Willson's fish and for Steve Kandra's fields have aggravated the rivalry between the Indian tribes living near the northern California coast and the irrigating farmers upstream along Oregon's arid southern border. The trouble, as farmers see it, came to a boiling point in 1997. That's the year coho salmon were accorded federal protection under the Endangered Species Act, which would entitle them to minimum flows of water. In 2001 tensions came to a dramatic head when the federal government shut off irrigation water to some 1,400 Klamath Reclamation Project farmers, including Kandra. The families felt singled out—"Farmers aren't used to being vilified," Kandra notes—and some responded with civil disobedience. They partially opened the irrigation canals' headgates in defiance of federal marshals and queued up for a symbolic bucket brigade through the streets of Klamath Falls, Oregon.

That summer the upper basin was a dry Dust Bowl flashback to *The Grapes of Wrath*. But by the following spring, reportedly thanks to Vice President Dick Cheney's behind-the-scenes intervention, the situation had reversed. In March 2002, Agriculture

Secretary Ann Veneman and Secretary of the Interior Gale Norton flew to Klamath Falls to open the valve into the main diversion canal and assure farmers they would have the water they needed. Matter settled.

Then came the sequel.

As Thomas Willson recounts, in September 2002, a vanguard of the fall salmon migration passed the coastal sandbar at Requa, California, and entered the mouth of the Klamath. The fish swam as far upstream as Blue Creek, a popular deep-pool gathering ground for their run up the river. Then, perhaps because the water in the slack river was so warm, they retreated back to the estuary. Rain in the Siskiyou Mountains cooled the river enough to encourage the fish to head back upstream, but when the weather turned sunny and hot, the fish, wearied by the false start and weakened by infections, didn't get far: At least 30,000 chinook salmon died in the lower 40 miles. Their carcasses carpeted the Klamath's banks in one of the largest adult fish die-offs in U.S. history.

The root causes of the massive fish kill remain disputed—there had been warmer temperatures and lower water levels, without disaster—but it certainly seemed to fulfill the dire prophesy of those who had opposed the opening of the floodgates and the constriction of river flows. Indian tribes and farmers and commercial ocean fishermen (who can have their seasons curtailed when salmon are scarce) confronted each other over flow rates and toxic algae, environmentalists insisted that farmers be evicted from leased land on Klamath wildlife refuges, and almost everyone squared off against Pacific Power, the company that owned the hydroelectric dams controlling the flow of the water. An epic American free-for-all erupted.

In the annals of father-to-son enterprises, the three-generation Kandra franchise may not boast the longevity of Willson & Co., but it can still be expressed in epic terms: Kandras have cultivated Klamath land ever since it became land. The two stretches of open field and farmyard homesteaded by Steve Kandra's grandfather and father look as solid as slab granite, but they bear liquid names: Lower

Klamath Lake to the west, and Tule Lake to the east. A little over a century ago, they were just that: expansive lakes.

Beginning in the early 1900s, in a mammoth engineering endeavor christened the Klamath Reclamation Project, much of the lake water was drained by the U.S. Reclamation Service to create new farms, more than 100,000 acres of them, and the new land was irrigated to make it arable. Hundreds of miles of canals and tunnels were built, and massive pumps installed to sluice water in and out. The "reclaimed" land in Tule Lake Basin was homesteaded, much of it by returning veterans of both World Wars whose names were drawn from a pickle jar; the farmers planted alfalfa, grain, potatoes, and onions on some of the most fertile soil in the West. Fertile because, as Kandra noted, shouting over the roar of his baler as he traced windrows of alfalfa in the headlights of his John Deere,

"down below us is a thousand feet of goose poop. It's old lake bottom. We're farming the top of a custard—you know how custard has a skin on it? We're on top of the skin."

The partition of the Klamath River was made concrete in 1918, when the California Oregon Power Co. (long known as Copco; later bought by Pacific Power) built the first of its big hydroelectric dams on the Klamath. Three other major dams followed, the farthest downstream being Iron Gate Dam, finished in 1962. Today the dams are the backbone of the power system that produces 750,000 megawatt hours for Pacific Power in an average year, enough to meet the electricity needs of 70,000 homes. It's especially useful power in that it releases no carbon emissions and can be turned on in an instant to supply peak needs.

The dams have long been a focus of local pride for the upriver communities, emblems of autonomy for a region that had always held

ADULT CHINOOK ENTERING THE KLAMATH RIVER EACH FALL
1978–2007, in thousands of fish

The Klamath's chinook salmon population is sharply reduced from historic highs. But today, the population varies dramatically year to year, reacting to an interplay of ocean conditions, the state of the river, and fishery management. Pacific Ocean conditions in the early 1990s led to several years of low counts. Poor water quality in the river could have led to the combination of infectious diseases, including Ichthyophthirius multifiliis, that resulted in a die-off of 30,500 adult chinook salmon in 2002.

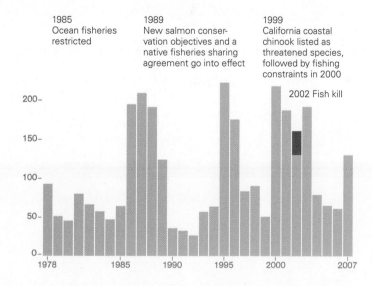

1985
Ocean fisheries restricted

1989
New salmon conservation objectives and a native fisheries sharing agreement go into effect

1999
California coastal chinook listed as threatened species, followed by fishing constraints in 2000

2002 Fish kill

MARTIN GAMACHE, NG STAFF (BOTH)
SOURCE: CHUCK TRACY, PACIFIC FISHERY MANAGEMENT COUNCIL

itself self-consciously apart. Residents call this stretch of far northern California and far southern Oregon the "State of Jefferson," and have on occasion discussed separating from their respective states and incorporating as a new state. Various efforts at statehood have faltered over the years, but Jefferson lives on as a code name for pugnacious patriotism. From the start, the local utility was a part of this independence. "I've heard that when they held an essay contest for the name of the new state they were going to form, the name that was suggested second to 'Jefferson' was 'Copcoland,'" says Toby Freeman, regional community manager with Pacific Power.

Whatever their utilitarian purpose, the dams effectively divided the river into two peoples, one of which lived off salmon, and the other of which never even saw one, since the dams obstructed the fish's upstream migrations. For the dams' opponents, the physical obstacle is

> The carcasses of 30,000 salmon carpeted the lower 40 miles of the Klamath River's banks in one of the largest fish die-offs in U.S. history.

only one of the ways the dams upset the Klamath's balance. Fishermen contend that the water impoundment alters the temperature and flow of river waters, encouraging fish diseases. In 2008, the Karuk Tribe released a report concluding that the cyanobacteria, commonly called blue-green algae, that bloom dramatically in the still summer waters behind Iron Gate are releasing toxins that could make fish and freshwater mussels unsafe to eat.

The issues in play over the Klamath's future are complex, but one prospect resides at the center of all the debate: removing the four hydroelectric dams. Advocates hope this might restore the river to its natural condition and allow migrating salmon an unobstructed path to headwater breeding grounds for the first time in a century. Demolition of the dams would also remove a symbolic barrier between the upriver and downriver human communities,

SPRING SNOWPACK TRENDS IN THE KLAMATH RIVER WATERSHED
1945–2002, at 32 monitoring stations

Percent change of the spring snowpack's water content

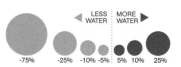

Mountain snowmelt delivers most of the water vital to fish and farms along the Klamath River. But since the mid-20th century, spring snowpack has declined significantly, with serious consequences. The situation is worst at lower elevations, where warming regional temperatures are reducing the territory where snow-pack can build up and persist late into the season. Scientists measure snowpack water content to estimate water availability during the spring and summer, when fish are on the move and irrigation demand peaks.

SOURCE: PHILIP MOTE, CLIMATE IMPACTS GROUP, UNIVERSITY OF WASHINGTON

but those parties haven't waited for dynamite to facilitate their convergence. For the past eight years, a group of affected parties—governmental, tribal, industrial, and private—has been convening over an endless series of conference tables in drab offices and motel meeting rooms, working its way through a cat's cradle of interlocking questions. If the talks succeed in resolving the Klamath conflict, the result will be historic. And if the dams are removed, notes Craig Tucker, Klamath Campaign coordinator for the Karuk Tribe, "this will be the largest dam removal ever on an American river. This can be a model for environmental cooperation."

Just in time, some might say. The perils to the nation's rivers are growing dramatically, as population growth and rising water usage overtax watersheds and deplete aquifers. In the western United States, that skyrocketing demand is on a crash course with the alarming effects of climate change. In response to warming temperatures, winters are bringing less and less snow to the American West, and snowpack is mother's milk to rivers like the Klamath. The Cascades and other Northwest mountains whose snowmelt feeds the river are the harbingers of what's to come elsewhere. Since the 1940s they have seen a significant decline in total snow accumulation because they are lower in elevation and so more susceptible to the region's rising temperatures than other western mountains. All of which makes the decisions over how to handle the competing needs for the Klamath's waters even more crucial. In coming decades, as governmental agencies turn increased attention to rescuing the world's riverine ecologies, they may cast an eye back to the way the small and relatively isolated communities of the Klamath River watershed negotiated their entrenched local issues and resolved historic antagonisms.

Especially since until this year, those issues seemed so intractable, and the antagonisms so fierce. Toby Freeman of Pacific Power, the company that would be responsible for the dams' removal, understands those antagonisms as well as anyone. Last year, asked for his forecast on the outcome of the river negotiations, he responded with bureaucratic cheer. "In the long run, I'm looking forward to a resolution that fully addresses the river's health while providing the best outcome for our customers," he said.

"In the short run," he added, "I'll be happy if no one gets shot."

Perhaps it's appropriate that the sources of the Klamath River begin in a geographic region known informally as the "blast zone." The blast in question was the eruption 7,700 years ago of Mount Mazama, one of the restive volcanic cones of the High Cascades, in southernmost Oregon. Klamath Indians explained the explosion as a battle between the sky god Skell and Llao, the deity of the underworld. Geologists describe it more technically. A series of eruptions blew much of Mazama's molten understory skyward; a mile-wide column of pumice, ash, and gas climbed into the upper stratosphere. As the 12 cubic miles of mountain and mantle fell back earthward, it draped 320 million acres in tephra—volcanic ash and rock—in a layer as thick as 20 feet. The remaining bulk of Mazama's summit collapsed (the mountain lost about a mile of elevation during the eruption), and the caldera filled partially with water, creating the con-summate natural tourist attraction, Crater Lake. To the lee of the crater stretched a vast new living desert of pumice, and, with time, a broad-based forest ecosystem dominated by bitterbrush, aspen, and lodgepole pine. Crater Lake locals liken a stroll through the blast zone to walking in kitty litter. The pale granular topsoil crunches underfoot and emits effusions of smoke-fine dust.

Coursing through the kitty-litter landscape are the tannin-stained streams that make up the headwaters of the Klamath River. They are fed by snowmelt from the Cascades, but much of that melt doesn't run downhill as surface water. Instead it soaks deep into the absorbent tephra and bubbles up as springs to feed the Williamson and Sprague Rivers, which run into Upper Klamath Lake. The water flows from there into Lake Ewauna, the official beginning of the Klamath River, which then flows into the Cascades and along the rugged Siskiyous. Its progress looks decidedly *(Continued on page 114)*

A RIVER UPSIDE DOWN

Instead of beginning like many rivers in remote mountains and flowing to an outlet on a heavily populated coast, the Klamath River starts in a well-peopled agricultural region where it's heavily tapped for irrigation. The 250-mile-long river arcs across a ten-million-acre-plus basin, flowing through hydroelectric dams in its upper and middle stretches before reaching wilder territory approaching the Pacific Ocean.

Portland
OREGON
Klamath Falls
Klamath River watershed
Arcata
CALIFORNIA
San Francisco
Sierra Nevada

Crescent City

YUROK INDIAN RESERVATION

REDWOOD

NATIONAL

AND STATE

PARKS

Requa
Klamath

Pecwan

Arcata

HOOPA VALLEY INDIAN RESERVATION

Hoopa

Weitchpec

Willow Creek

SISKIYOU

MOUNTAINS

Happy Camp

Seiad Valley

Condrey Mountain 7,112 ft 2,168 m

Somes Bar

Forks of Salmon

Marble Mountains

Salmon Mts.

Etna

Fort Jones

Yreka
Montague

5

MOUNTAIN

OREGON
CALIFORNIA

KLAMATH MOUNTAINS

Trinity Alps
Coffee Creek

Trinity Center

Weaverville

TRINITY DAM

LEWISTON DAM

Hayfork

Weed

97

Mt. Shasta

Mount Shasta 14,162 ft 4,317 m

Trinity Mountains

California Diversions
One of the Klamath's chief tributaries, the Trinity River is an important contributor to flow in the lower portion of the Klamath. Yet up to 90 percent of the Trinity's water is diverted at times to California for irrigation and navigation.

CLEAR CR. DIVERSION TUNNEL

Black Rock Mountain 7,792 ft 2,375 m

Redding

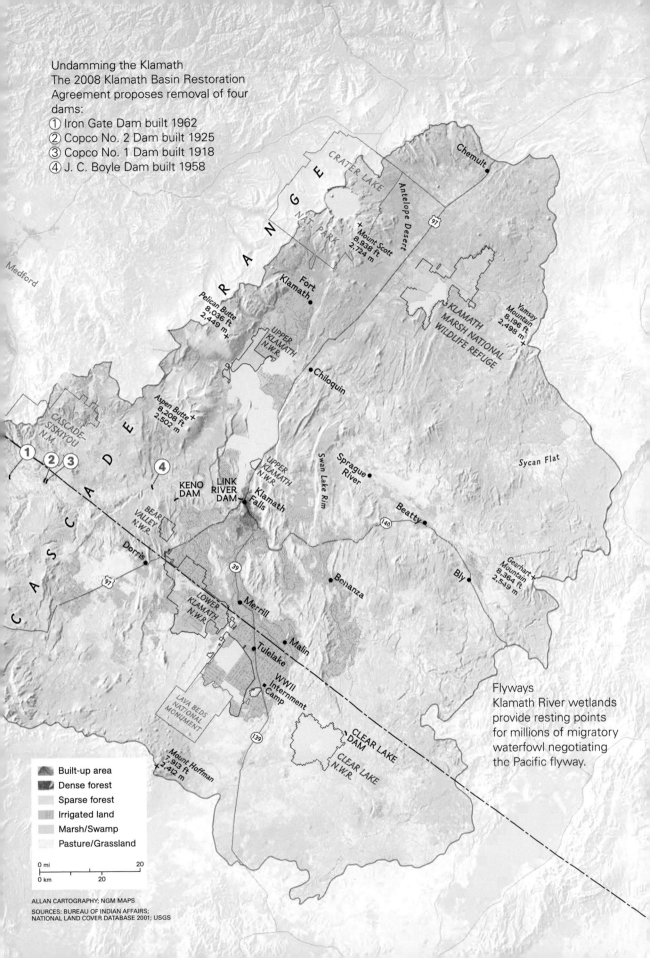

Undamming the Klamath
The 2008 Klamath Basin Restoration
Agreement proposes removal of four
dams:
① Iron Gate Dam built 1962
② Copco No. 2 Dam built 1925
③ Copco No. 1 Dam built 1918
④ J. C. Boyle Dam built 1958

Medford

CRATER LAKE

CASCADE RANGE

NAT. PARK

Chemult

Antelope Desert

97

+ Mount Scott
8,938 ft
2,724 m

Fort
Klamath

Pelican Butte
8,036 ft
2,449 m +

UPPER
KLAMATH
N.W.R.

Chiloquin

KLAMATH NATIONAL
MARSH
WILDLIFE REFUGE

Yamsay
Mountain
8,196 ft
2,498 m
+

Aspen Butte +
8,208 ft
2,502 m

UPPER
KLAMATH
N.W.R.

Swan Lake Rim

Sprague
River

Sycan Flat

CASCADE

① ② ③

④

KENO
DAM

LINK
RIVER
DAM

Klamath
Falls

Klamath
Falls

Beatty

140

BEAR
VALLEY
N.W.R.

Dorris

97

39

LOWER
KLAMATH
N.W.R.

Merrill

Bonanza

Bly

Gearhart +
Mountain
8,364 ft
2,549 m

Malin

Tulelake

WWII
Internment
Camp

LAVA BEDS
NATIONAL
MONUMENT

139

CLEAR LAKE
DAM

CLEAR
LAKE
N.W.R.

Flyways
Klamath River wetlands
provide resting points
for millions of migratory
waterfowl negotiating
the Pacific flyway.

Mount Hoffman
7,913 ft
2,412 m

Built-up area
Dense forest
Sparse forest
Irrigated land
Marsh/Swamp
Pasture/Grassland

0 mi   20

0 km   20

ALLAN CARTOGRAPHY; NGM MAPS

SOURCES: BUREAU OF INDIAN AFFAIRS;
NATIONAL LAND COVER DATABASE 2001; USGS

Crater Lake's 1,943-foot depths were created following volcanic eruptions 7,700 years ago; a lava outcrop juts from the northwest rim of the caldera. Snowmelt from the surrounding Cascade Mountains percolates through the land to the south, emerging as springs that feed streams flowing into Upper Klamath Lake.

(Continued from page 109) odd to anyone who's ever seen a…well, to anyone who's seen a river.

"It's a river upside down," Steve Pedery explained one summer afternoon, echoing a phrase often used to describe the Klamath. Pedery was speaking over the propeller roar of a Cessna. He is conservation director of Oregon Wild, an environmental organization, and the plane was courtesy of LightHawk, a group that provides overflights of ecological battlegrounds. "You know," Pedery said, as the lakes below us dwindled into a thin tinsel ribbon, "most rivers begin in the mountains, flow into farmland, and end up in a heavily industrialized urban port. The Klamath starts in farmland and flows through mountains that become wilder the closer you get to the coast."

The river is upended in another way. Most rivers begin pristine and wind up filthy. The Klamath gets "dirty" at its outset and becomes cleaner as it goes along. Even without the significant agricultural pollution feeding the profuse blue-green algae that skews the ecology on the upper Klamath, the river would have high levels of nutrients such as phosphorus, derived from the lake-bed soils of Upper Klamath Lake. Similarly, the warmth of the water may be exacerbated by dam impoundment and basking in farmers' fields, but the river was always naturally warmed by wide shallow lakes at its source. Only as the Klamath is joined along its descent by more traditional tributaries, the Trinity and the Scott and the Salmon, does it clean up and, at least temporarily, cool down.

The Klamath's upside-down design means it is exceptionally well poised to benefit from restoration. It has little of the industry and suburban development that clutter the shores of most American rivers. Most of its last 40 miles can only be visited by boat; no through road follows the river's course, and remote tribal villages like Pecwan, where Thomas Willson launches his fishing boat, are beyond the reach of electricity.

Winters are bringing less and less snow to the American West, and snowpack is mother's milk to western rivers like the Klamath.

If the dams are remade or removed, many experts agree that the Klamath could bounce back and become perhaps the healthiest big salmon river in the West. Maybe just as remarkable, saving the river could put an end to an unlovely slugfest among parties that historically, at least in the case of farmers (read pioneers) and Indians (read Indians), had been drawing each other's blood for a century and a half.

The Indians of the Klamath River watershed were among the very last Native Americans to be overrun by Manifest Destiny. A handful of tribes called the region home: the Modoc, Klamath, and Shasta Indians in the upper and middle basin, and the Karuk, Hoopa, and Yurok in the lower. The human onslaught that overtook them began in the mid-1800s; it would bring successive waves of settlers, gold miners, soldiers, loggers, farmers, and commercial fish canners. When the extraction of gold in the river's tributaries sent slurries of mud and tailings downstream, the native peoples got their first exposure to industrial pollution, and to the notion that an economy could be enriched by destroying its resources, instead of husbanding them. The Klamath River tribes were traders, and their currency, dentalium shells that tribespeople carried in oblong elk-horn purses, was valuable in relation to its scarcity. The shells were acquired from other tribes far to the north; unlike gold, they required no desecration to accumulate.

One hundred years after the gold rush, the lower Klamath welcomed an unusual visitor: Erik Erikson, the psychoanalyst who popularized the notion of the identity crisis. He'd come to study the Yurok Indian Tribe, whose worldview he described as "centripetal." Erikson meant that Yurok society was inwardly focused, a closed bell jar of a universe into which salmon and deer entered to sustain the tribe, but which the tribe's members never left. Their compass points were "away from the river" and "toward (Continued on page 120)

Behind the Klamath's oldest dam, a biologist samples soupy water (bottom) thickened by cyanobacteria in the Copco Lake. The toxic microorganism blooms when water heats up in summer. Some 30 miles east, irrigation channels outline fertile polygons on the drained bed of Lower Klamath Lake (top).

The Klamath winds through Oregon farmland, where Lincoln Gabriel rides out with his dogs to maintain irrigation ditches on 220 acres. Born in 1927, Gabriel has raised crops and run cattle here since his teens. "We ranchers paid for these ditches over 50 years," he says. "The government said we'd have the water we need."

Streams mix at the confluence of the Klamath and Salmon (at left) Rivers. More than a hundred miles downstream from the nearest dam—Iron Gate—the spot is more than just scenery to the Karuk Indians, who hold the Jump Dance, one of the High Dances of the World Renewal religion, on the riverbanks.

In the clear waters of the Salmon River, friends struggle for a grip on a greased watermelon during a party thrown by the Mid Klamath Watershed Council to celebrate watershed restoration and salmon conservation work.

(Continued from page 114) the river," as though the Klamath exerted an irresistible magnetic force that attracted much but let little go. The lower Klamath tribes share a religion known as World Renewal, which exalts nature's interconnectedness but sees that balance as precarious.

Leaf Hillman is vice chairman of the Karuk Tribe and a World Renewal priest whose family oversees a White Deer Dance, one of the rituals through which cosmic equilibrium is maintained. "This is a pretty unique place in the world," Hillman told me one day, kneeling by the entrance of a traditional sweathouse, a ten-foot-by-ten-foot-square structure topped with a gabled, wood-plank roof. The Klamath's currents burbled only yards away. When he was 13, Hillman was inducted into the Karuk priesthood and underwent a week of fasting and purification, sleeping by the fire in the sweat lodge and setting out each morning, dressed in deerskin and painted by an elder priest, on quests into the wild, learning the scripture of humanity's relationship with nature. Humans

have a responsibility to all other elements of nature, Hillman told me. "It's a reciprocal arrangement. We understand, and we know, that we owe our existence to the river."

The river was transportation and it was also sustenance, providing the willows used to make baskets and bringing the salmon and lamprey and trout that complemented acorns in the native diet. Back then, as the elders remember it, the fish were so thick that a person "could walk across the river on their backs," and salmon filled every belly. Traditionally, Karuk Tribe members each ate more than a pound of salmon a day, an intake that has dwindled in recent decades to under five pounds a year, with a commensurate surge in diabetes and heart disease.

With the disruption of the Indians' livelihood, and the river's inhabitants divided into mutually antagonistic communities, the Klamath faced the conundrum that stymies environmental efforts everywhere: Its problems were vast and expanding, but the communities that might solve them were too fragmented to mount a holistic response. Ironically, the historic bone of contention, the old Copco dams and what should happen to them, became the agent that would bring the warring parties together.

Four years after the massive salmon kill of 2002, the licenses for all four mid-Klamath dams came up for their 50-year renewal by the Federal Energy Regulatory Commission. In anticipation, Pacificorp, the parent company of Pacific Power, had begun meetings in 2000 with the Klamath area tribes, municipal governments, commercial fishermen, farmers, and environmental groups. The issues were daunting: What could replace the dams in providing Siskiyou County's tax base? If the dam removal succeeded in restoring salmon to the upper Klamath, what would happen if a farmer found an endangered coho in his irrigation canal; would he be shut down under the Endangered Species Act? Oregon Wild pushed hard to have farmers evicted from leased land on wildlife refuges; the Hoopa Valley Indian Tribe insisted that scientific studies be commissioned to verify that water-flow allotments would support the salmon. In 2006, after years of debate, and with talks expanding beyond dam removal and into such issues as tribal rights and river restoration, the group disbanded.

And then it reconvened, without Pacific Power involved, and with some of the more intransigent parties disinvited. Ron Cole, refuge manager for the Klamath Basin National Wildlife Refuges and, like Craig Tucker and Steve Kandra, a party to the talks, observed the turnaround. "The folks in this basin have never missed an opportunity to miss an opportunity, but I think they're tired of it," he told me. "This is considered ground zero for screwing up. But it can also be ground zero for success."

Last January the settlement parties announced the Klamath Basin Restoration Agreement, outlining options for saving the river. But negotiations with PacifiCorp over removal of

If the dams are remade or removed, the Klamath could bounce back and become perhaps the healthiest big salmon river in the West.

the dams—a key part of the plan—continue to drag on. Some entities, including the Hoopa tribe, remain unconvinced that their concerns have been addressed. And congressional action will be necessary to defray economic damages—implementing the agreement would cost hundreds of millions, perhaps billions, of dollars. Still, the most promising indication for success, and for the future of the Klamath, may have already taken place: the transformations within the individual river communities. The farmers took to heart their own observation of the Klamath's failing ecology. "People see that our farm inputs—oil, water, fertilizer—aren't infinite, like they seemed to be 20 years ago," Klamath farmer John Anderson said. The Andersons responded by shifting crops and refining their irrigation methods. Other farmers, like the Kandras, installed new pivot irrigators that are stingier with water than the old, crude field-flooding methods. In the Tule Lake Basin, farmers have also been rotating their fallow fields into lake and marsh as part of a U.S. Fish and Wildlife program called Walking Wetlands. The rotation is heralded as good for wildlife and for agriculture: It provides bulrush sanctuary to migratory birds while replenishing the land so that it is more productive when it again goes under cultivation.

Even such mutually beneficial arrangements required a laying down of old suspicions, noted Ron Cole, as he marched with wildlife biologist Dave Mauser through a soon-to-be-flooded field, wearing his Department of the Interior greens. "Years ago, there's no way we would be standing in these uniforms in this private field," he marveled.

For their part, the lower Klamath Indians have had to break out of a tradition of secretiveness—born of the time when they fished and worshipped only at night and spent their days hiding in caves, trying to stay invisible. Now they must join in a boisterous debate with people they've never been able to trust.

A Native American boy snuggles up to a brace of chinook salmon his family caught in Requa, California. Diverse communities depend on the river for sustenance—and often their needs clash. "We've all got to let go of hard feelings," farmer Scott Seus says, "and try to find a common way ahead."

"In my mind that's the very thing that saved us, our ability to blend in. But now that strategy has to change," Leaf Hillman explained, "because if we continue to blend in and not be noticed, that will spell our doom."

Meanwhile, the farmers upstream are sounding like nothing so much as World Renewal converts, proselytizing the community of all nature. "As a man of faith," Steve Kandra said, "I think the water crisis was God saying you guys gotta figure it out, because you're related to each other. You guys better figure it out. Well, us folks that are here on the ground, we're working darn hard to save these communities. I don't think anyone is going to accept elimination of one community over the other."

"What I think has evolved is that people are looking out for the other guy's back, not just their own anymore," Ron Cole observed. "Just a little. The families up here, they never felt connected with this river. Now they do. They feel they're river people too."

Discussion Questions

- How can the installation of dams affect water quality, agriculture and fish populations? Do you think the ecology of rivers can bounce back if they are removed, or is the loss of electricity and agricultural support is too great a threat to the surrounding population's way of life?

- Explain how the Klamath River System is different from many other river systems and why this makes its potential for restoration so much more appealing for fishermen, conservationists and resource managers.

- How can programs like the "Walking Wetlands Program" help people from different backgrounds see each other's point of view and bring them closer together on larger issues like dam removal? Do you think programs like this should be encouraged in other regions, or do they only apply to areas with these particular characteristics?

- While much progress has been made in bringing the sides together over potential restoration projects, the article cites many daunting issues that still remain. For example: What could replace the dams in providing a tax base? If the dam removal succeeded in restoring salmon to the upper Klamath, what would happen if a famer found an endangered coho in his irrigation canal; would he be shut down under the Endangered Species Act? What is your opinion on how issues like these should be resolved? Are there any other major issues that you think need to be addressed?

Map Literacy

Use the map and figures contained in this article to explain the following:

- Describe how the areas surrounding the Klamath River are used from the headwaters moving all the way down towards the ocean.

- Find the dams that are under consideration for removal under the Klamath Basin Restoration Agreement. How do you think this will affect water flows and usage above and below their current location?

- How would you describe the consistency of annual Chinook populations entering the Klamath River? What factors do you think are most responsible for this?

- Describe how snowpack water content has changed in the last 50 years. What affect does this have on water availability during the spring and summer months? How do these changes affect surrounding populations?

DRYING OF THE WEST

Drying of the West documents some of the changes that the U.S. Southwest is experiencing and describes how they these changes are affecting both the ecology and the regional population centers. It gives a historical perspective on why the people were so unprepared for recent droughts, and it looks forward to an uncertain future. The article discusses potential future projects that may help alleviate some of the current issues but also poses some important questions about equity, water use efficiency and how to judiciously plan for living in a drier climate.

When reading this article, you should focus on:

- How scientists use tree ring data to learn about past climates and what this data tells us about the frequency and degree of past droughts in the Southwest,
- What affect El Niño and La Niña have on the weather and climate of the Southwest,
- How forests are affected by climatic changes in the region, and
- What are some of the options that city planners and resource managers have to deal with changing water levels?

In drought-parched Los Padres National Forest in southern California, a heli-tanker douses a hot spot in the huge Zaca fire that erupted in July 2007, scorching 240,000 acres. Years of sparse rain primed the region for the second largest fire in California history.

DRYING OF
THE WEST

By Robert Kunzig
Photographs by Vincent Laforet

Well water allows the lush greens and fairways of the Primm Valley Golf Club to flourish in the Mojave Desert—a bone-dry blast furnace where only hardy desert plants normally survive. Though limits have been imposed, golf courses in nearby southern Nevada still use 8 percent of the region's water.

Lake Powell's "bathtub ring"—a residue from water immersion—records how far the water level has fallen in the giant reservoir. Inflow from the Colo-rado River has been below average every year but one since 1999, when Powell was last full. It's now below 50 percent capacity and dropping.

THE AMERICAN WEST WAS WON BY WATER MANAGEMENT. WHAT HAPPENS WHEN THERE'S NO WATER LEFT TO MANAGE?

When provided with continuous nourishment, trees, like people, grow complacent.

Tree-ring scientists use the word to describe trees like those on the floor of the Colorado River Valley, whose roots tap into thick reservoirs of moist soil. Complacent trees aren't much use for learning about climate history, because they pack on wide new rings of wood even in dry years. To find trees that feel the same climatic pulses as the river, trees whose rings widen and narrow from year to year with the river itself, scientists have to climb up the steep, rocky slopes above the valley and look for gnarled, ugly trees, the kind that loggers ignore. For some reason such "sensitive" trees seem to live longer than the complacent ones. "Maybe you can get too much of a good thing," says Dave Meko.

Meko, a scientist at the Laboratory of Tree-Ring Research at the University of Arizona, has been studying the climate history of the western United States for decades. Tree-ring fieldwork is hardly expensive—you need a device called an increment borer to drill into the trees, you need plastic straws (available in a pinch from McDonald's) to store

Maybe you can get too much of a good thing,

the pencil-thin cores you've extracted from bark to pith, and you need gas, food, and lodging. But during the relatively wet 1980s and early '90s, Meko found it difficult to raise even the modest funds needed for his work. "You don't generate interest to study drought unless you're in a drought," he says. "You really need a catastrophe to get people's attention," adds colleague Connie Woodhouse.

Then, in 2002, the third dry year in a row and the driest on record in many parts of the Southwest, the flow in the Colorado fell to a quarter of its long-term average. That got people's attention.

The Colorado supplies 30 million people in seven states and Mexico with water. Denver, Las Vegas, Phoenix, Tucson, Los Angeles, and San Diego all depend on it, and starting this year so will Albuquerque. It irrigates four million acres of farmland, much of which would otherwise be desert, but which now produces billions of dollars' worth of crops. Gauges first installed in the

Adapted from "Drying of the West" by Robert Kunzig: National Geographic Magazine, February 2008.

19th century provide a measure of the flow of the river in acre-feet, one acre-foot being a foot of water spread over an acre, or about 326,000 gallons. Today the operation of the pharaonic infrastructure that taps the Colorado—the dams and reservoirs and pipelines and aqueducts—is based entirely on data from those gauges. In 2002 water managers all along the river began to wonder whether that century of data gave them a full appreciation of the river's eccentricities. With the lawns dying in Denver, a water manager there asked Woodhouse: How often has it been this dry?

Over the next few years Woodhouse, Meko, and some colleagues hunted down and cored the oldest drought-sensitive trees they could find growing in the upper Colorado basin, both living and dead. Wood takes a long time to rot in a dry climate; in Harmon Canyon in eastern Utah, Meko found one Douglas fir log that had laid down its first ring as a sapling in 323 b.c. That was an extreme case, but the scientists still collected enough old wood to push their estimates of annual variations in the flow of the Colorado back deep into the Middle Ages. The results came out last spring. They showed that the Colorado has not always been as generous as it was throughout the 20th century.

The California Department of Water Resources, which had funded some of the research, published the results as an illustrated poster. Beneath a series of stock southwestern postcard shots, the spiky trace of tree-ring data oscillates nervously across the page, from a.d. 762 on the left to 2005 on the right. One photo shows the Hoover Dam, water gushing from its outlets. When the dam was being planned in the 1920s to deliver river water to the farms of the Imperial Valley and the nascent sprawl of Los Angeles, the West, according to the tree rings, was in one of the wettest quarter

The wet 20th century, the wettest of the past millennium, the century when Americans built an incredible civilization in the desert, is over.

centuries of the past millennium. Another photo shows the booming skyline of San Diego, which doubled its population between 1970 and 2000—again, an exceptionally wet period along the river. But toward the far left of the poster, there is a picture of Spruce Tree House, one of the spectacular cliff dwellings at Mesa Verde National Park in southwestern Colorado, a pueblo site abandoned by the Anasazi at the end of the 13th century. Underneath the photo, the graph reveals that the Anasazi disappeared in a time of exceptional drought and low flow in the river.

In fact, the tree rings testified that in the centuries before Europeans settled the Southwest, the Colorado basin repeatedly experienced droughts more severe and protracted than any since then. During one 13-year megadrought in the 12th century, the flow in the river averaged around 12 million acre-feet, 80 percent of the average flow during the 20th century and considerably less than is taken out of it for human use today. Such a flow today would mean serious shortages, and serious water wars. "The Colorado River at 12 million acre-feet would be real ugly," says one water manager.

Unfortunately, global warming could make things even uglier. Last April, a month before Meko and Woodhouse published their latest results, a comprehensive study of climate models reported in Science predicted the South-west's gradual descent into persistent Dust Bowl conditions by mid-century. Researchers at the National Oceanic and Atmospheric Administration (NOAA), meanwhile, have used some of the same models to project Colorado streamflow. In their simulations, which have been confirmed by others, the river never emerges from the current drought. Before mid-century, its flow falls to seven million acre-feet—around half the amount consumed today.

The wet 20th century, the wettest of the past millennium, the century when Americans built an incredible civilization in the desert, is over. Trees in the West are adjusting to the change, and not just in the width of their annual rings: In the recent drought they have been dying off and burning in wildfires at an unprecedented rate. For most people in the region, the news hasn't quite sunk in. Between 2000 and 2006 the seven states of the Colorado basin added five million people, a 10 percent population increase. Subdivisions continue to sprout in the desert, farther and farther from the cities whose own water supply is uncertain. Water managers are facing up to hard times ahead. "I look at the turn of the century as the defining moment when the New West began," says Pat Mulroy, head of the Southern Nevada Water Authority. "It's like the impact of global warming fell on us overnight."

In July 2007 a few dozen climate specialists gathered at Columbia University's Lamont-Doherty Earth Observatory to discuss the past and future of the world's drylands, especially the Southwest. Between sessions they took coffee and lunch outside, on a large sloping lawn above the Hudson River, which gathers as much water as the Colorado from a drainage area just over a twentieth the size. It was overcast and pleasantly cool for summer in New York. Phoenix was on its way to setting a record of 32 days in a single year with temperatures above 110 degrees. A scientist who had flown in from the West Coast reported that he had seen wildfires burning all over Nevada from his airplane window.

On the first morning, much of the talk was about medieval megadroughts. Scott Stine of California State University, East Bay, presented vivid evidence that they had extended beyond the Colorado River basin, well into California. Stine works in and around the Sierra Nevada, whose snows are the largest source of water for that heavily populated state. Some of the runoff drains into Mono Lake on the eastern flank of the Sierra. After Los Angeles began diverting the streams that feed Mono Lake in the 1940s, the lake's water level dropped 45 vertical feet.

In the late 1970s, tramping across the newly exposed shorelines, Stine found dozens of tree stumps, mostly cottonwood and Jeffrey pine, rooted in place. They were gnarled and ancient looking and encased in tufa—a whitish gray calcium carbonate crust that precipitates from the briny water of the lake. Clearly the trees had grown when a severe and long-lasting drought had lowered the lake and exposed the land where they had taken root; they had died when a return to a wetter climate in the Sierra Nevada caused the lake to drown them. Their rooted remains were now exposed because Los Angeles had drawn the lake down.

Stine found drowned stumps in many other places in the Sierra Nevada. They all fell into two distinct generations, corresponding to two distinct droughts. The first had begun sometime before 900 and lasted over two centuries. There followed several extremely wet decades, not unlike those of the early 20th century. Then the next epic drought kicked in for 150 years, ending around 1350. Stine estimates that the runoff into Sierran lakes during the droughts must have been less than 60 percent of the modern average, and it may have been as low as 25 percent, for decades at a time. "What we have come to consider normal is profoundly wet," Stine said. "We're kidding ourselves if we think that's going to continue, with or without global warming."

No one is sure what caused the medieval megadroughts. Today Southwestern droughts follow the rhythm of La Niña, a periodic cooling of the eastern equatorial Pacific. La Niña alternates every few years with its warm twin, El Niño, and both make weather waves around the globe. A La Niña cooling of less than a degree Celsius was enough to trigger the recent drought, in part because it shifted the jet stream and the track (Continued on page 136)

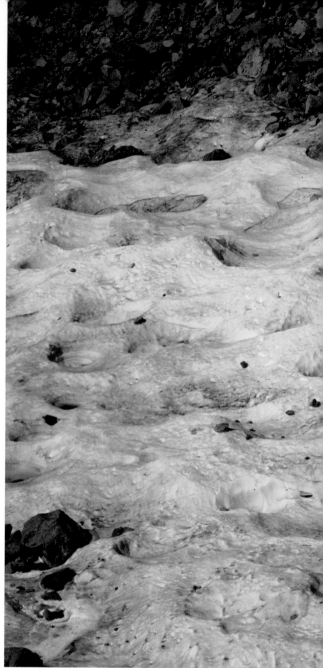

As the West dries out, the landscape is transformed. Without cold winters to kill off their larvae, mountain pine beetles infest up to 90 percent of lodgepole pines in Colorado forests, like this one near Granby (above left). The dead trees raise the risk of wildfires. In much of the West warmer, drier winters have reduced snowpack, a crucial water source. On California's Mount Shasta (above) a hiker traverses a snow patch diminished by milder temperatures.

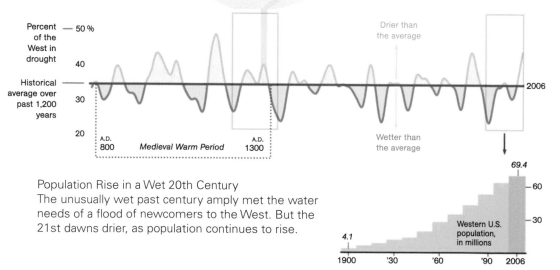

READING THE RINGS

Climate patterns of centuries past can be tracked in a tree's annual growth rings: Dry years produce thinner bands than wet years. A wedge from a Douglas fir log (above), collected in Utah's Harmon Canyon, holds the precipitation record of the upper Colorado River Basin from the 10th to 17th centuries a.d. The enlarged section below highlights a decade-long drought in the late 1200s that likely drove the Anasazi from Mesa Verde.

PHOTO: LABORATORY Of TREE-RING RESEARCH, UNIVERSITY OF ARIZONA. LENGTH OF WEDGE SHOWN IS 12 inches.

Less water, thinner rings

A.D. **1196**

A.D. **1322**

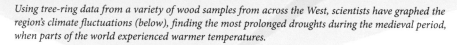

Using tree-ring data from a variety of wood samples from across the West, scientists have graphed the region's climate fluctuations (below), finding the most prolonged droughts during the medieval period, when parts of the world experienced warmer temperatures.

Percent of the West in drought

— 50 %

40

Historical average over past 1,200 years

30

20

Drier than the average

Wetter than the average

A.D. 800 — *Medieval Warm Period* — A.D. 1300

2006

69.4

Population Rise in a Wet 20th Century
The unusually wet past century amply met the water needs of a flood of newcomers to the West. But the 21st dawns drier, as population continues to rise.

4.1

Western U.S. population, in millions

— 60

— 30

1900   '30   '60   '90   2006

CHART ABOVE IS SMOOTHED USING A 50-YEAR MOVING AVERAGE. DATA: EDWARD COOK, TREE-RING LABORATORY, LAMONT-DOHERTY EARTH OBSERVATORY, COLUMBIA UNIVERSITY; U.S. CENSUS BUREAU. REPORTING AND GRAPHICS BY TOM ZELLER, JR., NGM ART

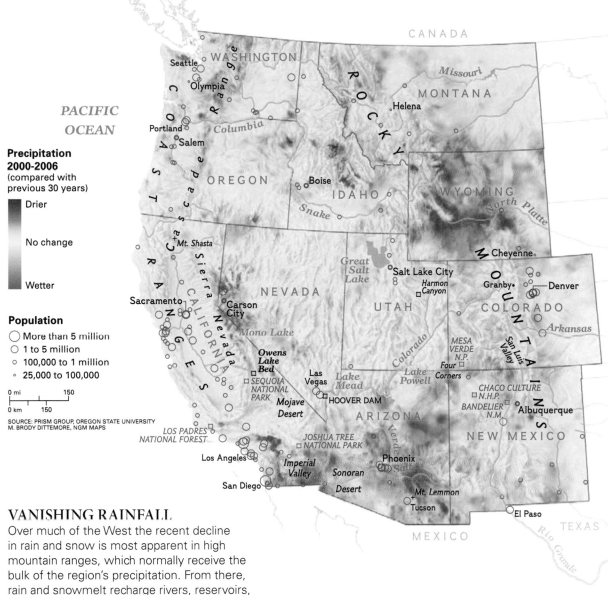

Precipitation 2000-2006
(compared with previous 30 years)

Drier

No change

Wetter

Population

○ More than 5 million

○ 1 to 5 million

○ 100,000 to 1 million

· 25,000 to 100,000

0 mi ——— 150

0 km ——— 150

SOURCE: PRISM GROUP, OREGON STATE UNIVERSITY
M. BRODY DITTEMORE, NGM MAPS

VANISHING RAINFALL

Over much of the West the recent decline in rain and snow is most apparent in high mountain ranges, which normally receive the bulk of the region's precipitation. From there, rain and snowmelt recharge rivers, reservoirs, and aquifers.

ONCE AND FUTURE DROUGHT

The West is naturally dry, but just how dry it can get is only now being understood. In contrast to the 20th century, revealed by tree rings as the wettest of the past millennium, an era called the Medieval Warm Period was dominated by deep droughts. Those megadroughts lowered the flow of the Colorado River to less than the volume currently drawn from it by 30 million people from Los Angeles to Denver for consumption and irrigation.

Natural cycles of drought in the West, especially the Southwest, are thought to be triggered mainly by the Pacific Ocean phenomenon called La Niña, a pulse of cooler equatorial water that periodically shifts the jet stream and its winter storms to the north. With the effects of La Niña expected to be compounded in coming decades by global warming, the politics of sharing the Colorado—and all western water resources—will only intensify.

(Continued from page 131) of the winter storms northward, out of the Southwest. Richard Seager, of La-mont, and his colleagues have shown that all the western droughts in the historical record, including the Dust Bowl, can be explained by small but unusually persistent La Niñas. Though the evidence is slimmer, Seager thinks the medi-eval megadroughts too may have been caused by the tropical Pacific see-saw getting stuck in something like a perpetual La Niña.

The future, though, won't be governed by that kind of natural fluctuation alone. Thanks to our emissions of greenhouse gases, it will be subject as well to a global one-way trend toward higher temperatures. In one talk at Lamont, climate theorist Isaac Held, from NOAA's Geophysical Fluid Dynamics Laboratory in Princeton, gave two reasons why global warming seems almost certain to make the drylands drier. Both have to do with an atmospheric circulation pattern called Hadley cells. At the Equator, warm, moist air rises, cools, sheds its moisture in tropical downpours, then spreads toward both Poles. In the subtropics, at latitudes of about 30 degrees, the dry air descends to the surface, where it sucks up moisture, creating the world's deserts—the Sahara, the deserts of Australia, and the arid lands of the Southwest. Surface winds export the moisture out of the dry subtropics to temperate and tropical latitudes. Global warming will intensify the whole process. The upshot is, the dry regions will get drier, and the wet regions will get wetter. "That's it," said Held. "There's nothing subtle here. Why do we need climate models to tell us that? Well, we really don't."

A second, subtler effect amplifies the drying. As the planet warms, the poleward edge of the Hadley cells, where the deserts are, expands a couple of degrees latitude farther toward each Pole. No one really knows what causes this effect—but nearly all climate models predict it, making it what modelers call a robust result. Because the Southwest is right on the northern edge of the dry zone, a northward shift will plunge the region deeper into aridity.

As the meeting neared its close, Held and Seager stood out on the lawn, discussing Hadley cells and related matters through mouthfuls of coffee and doughnuts. The two men had lately become collaborators, and a few months before had published with colleagues the sobering Science paper analyzing the results of 19 different simulations done by climate modeling groups around the world. They then averaged all these results into an "ensemble."

The ensemble shows precipitation in the Southwest steadily declining over the next few decades, until by mid-century, Dust Bowl conditions are the norm. It does not show the Pacific locked in a perpetual La Niña. Rather, La Niñas would continue to happen as they do today (the present one is expected to continue at least through the winter of 2008), but against a background state that is more profoundly arid. According to the ensemble model, the descent into that state may already have started.

People are not yet suffering, but trees are. Forests in the West are dying, most impressively by burning. The damage done by wildfires in the U.S., the vast majority of them in the West, has soared since the late 1980s. In 2006 nearly ten million acres were destroyed—an all-time record matched the very next year. With temperatures in the region up four degrees F over the past 30 years, spring is coming sooner to the western mountains. The snowpack—already diminished by drought—melts earlier in the year, drying the land and giving the wildfire season a jump start. As hotter summers encroach on autumn, the fires are ending later as well.

The fires are not only more frequent; they are also hotter and more damaging—though not entirely because of climate change. According to Tom Swetnam, director of the University of Arizona tree-ring lab, the root cause is the government's policy, adopted early in the 20th century, of trying to extinguish all wildfires. By studying sections cut from dead, thousand-year-old giant sequoias in the Sierra Nevada and from ponderosa pines all over Arizona and New Mexico, Swetnam discovered that most southwestern forests have always burned often—but at low intensity, with flames just a few feet high that raced through the grasses and the needles on the forest floor. The typical tree bears the marks of many such events, black scars where flames ate through the bark and perhaps even took a deep wedge out of the tree, but left it alive to heal its wound with new growth. Suppressing those natural fires has produced denser forests, with flammable litter piled up on the floor, and thickets of shrubs and young trees that act as fire ladders. When fires start now, they don't stay on the ground—they shoot up those ladders to the crowns of the trees. They blow thousand-acre holes in the forest and send mushroom clouds into the air.

One day last summer, Swetnam took a few visitors up Mount Lemmon, just north of Tucson, to see what the aftermath of such events looks like. In May 2002 the Bullock fire roared up the northeast slope of Mount Lemmon, consuming 30,000 acres. Firefighters stopped it at the Catalina Highway, protecting the village of Summerhaven. But the very next year, the Aspen fire started on the slope just below the village, destroying nearly half of the 700-odd houses in Summerhaven and burning 85,000 acres, all the way down to the outskirts of Tucson. The entire mountainside beyond the village remains covered with the gray skeletons of ponderosa pines, like one big blast zone. "Ponderosa pine is not adapted to these crown fires," Swetnam said, contemplating the site from the scenic overlook above the village. "It has heavy, wingless seeds that don't go very far. When you get a large hole like this, it will take hundreds of years to fill in from the edges."

> Sequoias may not survive in Sequoia National Park. What do you do? Do you irrigate these things? Or do you let a 2,000-year-old tree die?

Mount Lemmon's forests are also experiencing a slower, broader change. The Catalina Highway starts out flat, at an altitude of 2,500 feet in the Sonoran Desert, with its saguaros and strip malls. As the road leaves the last of Tucson behind, it climbs steeply through the whole range of southwestern woodland ecosystems—first scrub oak, then piñon and juniper, then ponderosa pine and other conifers, until finally, after less than an hour and a climb of 7,000 feet, you reach the spruce and fir trees on the cool peak. There is a small ski area there, the southernmost in the United States, and its days are certainly numbered.

As Swetnam explained, the mountain is one of an archipelago of "sky islands" spread across southeastern Arizona, New Mexico, Texas, and into Mexico—mountains isolated from one another by a sea of desert or grassland. Like isles in the ocean, these islands are populated in part by endemics—species that live nowhere else. The sky-island endemics are cool- and wet-loving species that have taken refuge on the mountaintops since the last ice age. They are things like the corkbark fir, or the endangered red squirrel that lives only on nearby Mount Graham. Their future is as bleak as that of the ski area. "They'll be picked off the top," said Swetnam. "The islands are shrinking. The aridity is advancing upslope."

All over the Southwest, a wholesale change in the landscape is under way. Piñons and scrubbier, more drought-resistant junipers have long been partners in the low woodlands that clothe much of the region. But the piñons are dying off. From 2002 to 2004, 2.5 million acres turned to rust in the Four Corners region alone. The immediate cause of death was often bark beetles, which are also devastating other conifers. The Forest Service estimates that in 2003, beetles infested 14 million (Continued on page 142)

A backyard pool is the perfect remedy to Phoenix's brutal heat. The drought has yet to put a serious dent in the lifestyles of millions who moved to the region in recent decades. Water from aquifers and the Colorado and Salt Rivers has so far allowed Phoenix to avoid restrictions imposed in other desert cities.

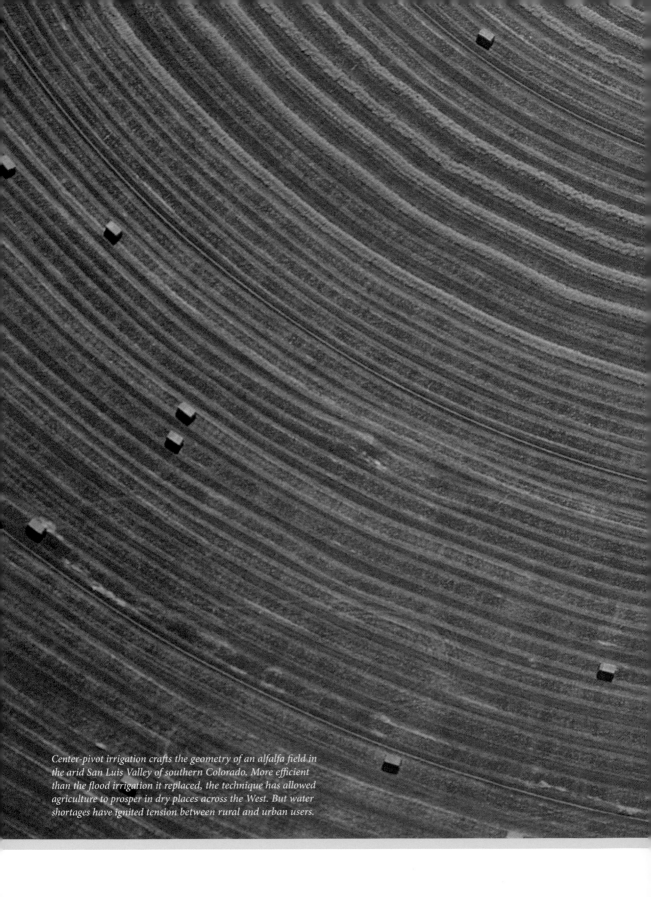

Center-pivot irrigation crafts the geometry of an alfalfa field in the arid San Luis Valley of southern Colorado. More efficient than the flood irrigation it replaced, the technique has allowed agriculture to prosper in dry places across the West. But water shortages have ignited tension between rural and urban users.

(*Continued from page 137*) acres of piñon, ponderosa, lodgepole pine, and Douglas fir in the American West.

Bark beetles tend to attack trees that are already stressed or dying from drought. "They can smell it," says Craig Allen, a landscape ecologist at Bandelier National Monument in the Jemez Mountains of New Mexico. Global climate change may be permanently teasing the piñons and junipers apart, and replacing piñon-juniper woodland with something new. At Bandelier, Allen has observed that junipers, along with shrubs such as wavyleaf oak and mountain mahogany, now dominate the beetle-ravaged landscape: pockets of green gradually spreading beneath a shroud of dead piñons.

Just as there are global climate models, there are global models that forecast how vegetation will change as the climate warms. They predict that on roughly half of Earth's surface, something different will be growing in 2100 than is growing there now. The models are not good, however, at projecting what scientists call "transient dynamics"—the damage done by droughts, fires, and beetle infestations that will actually accomplish the transformation. Large trees cannot simply migrate to higher latitudes and altitudes; they are rooted to the spot. "What happens to what's there now?" Allen wonders. "Stuff dies quicker than it grows."

Over the next few decades, Allen predicts, people in the Southwest will be seeing a lot of death in the old landscapes while waiting for the new ones to be born. "This is a dilemma for the Park Service," he says. "The projections are that Joshua trees may not survive in Joshua Tree National Park. Sequoias may not survive in Sequoia National Park. What do you do? Do you irrigate these things? Or do you let a 2,000-year-old tree die?"

While the trees die, the subdivisions proliferate. "Our job was to entice people to move to the West, and we did a darn good job," says Terry Fulp, who manages water releases at Hoover Dam. The federal Bureau of Reclamation built the dam in the 1930s primarily to supply the vegetable farms of the Imperial Valley and only secondarily to supply the residents of Los Angeles. Farmers had first claim to the water—they still do—but there was plenty to go around. "At Lake Mead, we basically gave the water away," says Fulp. "At the time, it made perfect sense. There was no one out here." After Reclamation built Hoover and the other big dams, more people came to the desert than anyone ever expected. Few of them are farmers anymore, and farming, crucial as it is to human welfare, is now a small part of the economy. But it still uses around three-quarters of the water in the Colorado River and elsewhere in the Southwest.

In the wet 1920s, as the dam was being planned, seven states drew up the Colorado River Compact to divvy up 15 million acre-feet of its water. California, Nevada, and Arizona—the so-called Lower Basin states—would get half, plus any surplus from the Upper Basin states of Wyoming, Colorado, New Mexico, and Utah. The compact also acknowledged Mexico's rights to the water. Surpluses were almost always on hand, because the Upper Basin states have never fully used the 7.5 million acre-feet they are entitled to under the compact. They are only entitled to use it, in fact, if in so doing they don't prevent the Lower Basin states from getting their 7.5 million—the compact is unfair that way. But in the wet 20th century, it didn't seem to matter.

In 1999 both Lake Mead and Lake Powell—created in 1963 upstream of Lake Mead to ensure that the Upper Basin would have enough water even in drought years to meet its obligation to the Lower Basin—were nearly full, with 50 million acre-feet between them. Two years later, representatives of the states in the basin completed long and difficult negotiations with the Bureau of Reclamation on new guidelines for dividing up the surpluses from Lake Mead. Then came the drought. Both lakes are now only half full. "Those guidelines are almost a joke now," says the Southern Nevada Water Authority's Pat Mulroy. "All of a sudden, seven states that had spent years in surplus discussions had to turn on a dime and start discussing shortages."

Mulroy, a crisp, tanned, fiftysomething blonde with a tailored look and a forceful personality, has run the Las Vegas water district since 1989. During that time she has watched the area's population growth consistently outstrip demographic projection. The population is almost two million now, having grown by 25 percent during the drought years; Mulroy is convinced it will go to three million. Before the drought, she and her colleagues nevertheless thought their water supply, 90 percent of it from Lake Mead, was safe for 50 years. In 2002 they were celebrating the opening of a second water intake from Lake Mead, 50 feet lower than the old one, which more than doubled their pumping capacity. Now they are scrambling to insert a third "straw" even deeper into the sinking lake. Las Vegas is also trying to reduce its dependence on the Colorado. The SNWA is exercising water rights and buying up ranches in the east-central part of the state. It plans to sink wells and tap groundwater there and pump as much as 200,000 acre-feet of it through a 250-mile pipeline to the city. There is considerable local opposition, of course, and an environmental impact statement must be prepared—but there is "zero chance," Mulroy says grimly, that the pipeline won't be built.

Other southwestern cities are also realizing their vulnerability to drought. Phoenix, hellish as it is in summer and bisected by the dry bed of the Salt River, is better off than most—for the moment. "In 2002 Phoenix was virtually the only city in the Southwest that had no mandatory restrictions," says Charlie Ester, water resources manager at the Salt River Project in

The West was built by dreamers. As the climate that underpinned that expansive vision vanishes, the vision needed to replace it has not yet emerged.

Phoenix. "We didn't need them." Phoenix pumps groundwater whenever it needs to, though it is under a state mandate to stop depleting the aquifer. And it gets a little over a third of its water from the Colorado River via the Central Arizona Project, a 336-mile-long canal. But the Salt River remains its biggest source. The riverbed is dry in the city because the SRP has half a dozen dams in the mountains north and east of the city, which convert the Salt and its tributary, the Verde, into chains of terraced lakes.

Phoenix would thus seem to possess that holy grail of water managers: a diversified portfolio. But Ester was still disconcerted to see his lake levels dropping in the drought, until they were less than half full. After he called the tree-ring lab, Dave Meko and climatologist Katie Hirschboeck looked into the tree-ring records for the Salt and Verde Rivers' watersheds.

"They found they were virtually identical," Ester says. "There were only three years out of 800 where the Colorado was wet and the Salt was dry or vice versa. What that means is, if we have a bad drought in Arizona, and the Salt dries up, we can't rely on the Colorado to bail us out. So what are we going to do? Well, we're going to hurt. Or move."

Since the Hoover Dam was built, there has never been a water shortage on the Colorado, never a day when there was simply not enough water in Lake Mead to meet all the downstream allocations. Drought, and a realistic understanding of the past, have made such a day seem more imminent. Under the pressure of the drought, the seven Colorado basin states have agreed for the first time on how to share prospective shortages. Arizona will bear almost all the pain at first, because the Central Arizona Project, which came on line in 1993, has junior rights. Nevada will lose only a small percentage of its allotment. *(Continued on page 146)*

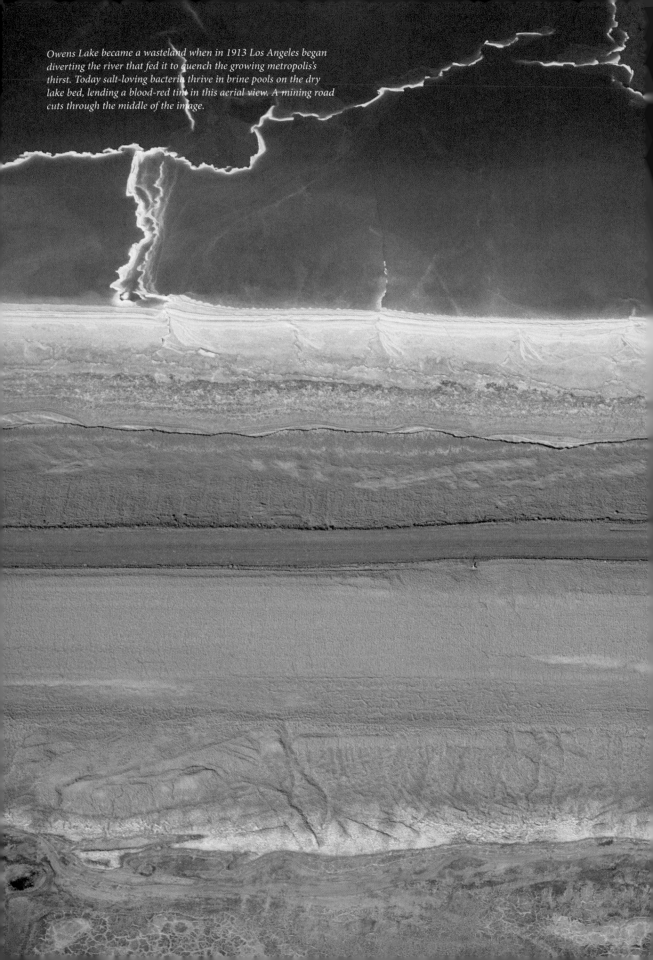

Owens Lake became a wasteland when in 1913 Los Angeles began diverting the river that fed it to quench the growing metropolis's thirst. Today salt-loving bacteria thrive in brine pools on the dry lake bed, lending a blood-red tint in this aerial view. A mining road cuts through the middle of the image.

A viable desert home during a long wet spell may be uninhabitable when the rains stop. The ancient Anasazi created a flourishing culture in New Mexico's Chaco Canyon, epitomized by Pueblo Bonito. Then prolonged drought hit the region in 1130. By the time it ended 30 years later, the Anasazi were gone. Sprawling cities in the present-day

(Continued from page 143) Meanwhile California would give up nothing, at least until Lake Mead falls below 1,025 feet, nearly 200 feet below "full pool." At that point, negotiations would resume. According to Bureau of Reclamation calculations, a return of the 12th-century drought would force Lake Mead well below that level, perhaps even to "dead pool" at 895 feet—the level at which water no longer flows out of the lake without pumping. Reclamation officials consider this extremely unlikely. But their calculations do not take into account the impact of global warming.

Every utility in the Southwest now preaches conservation and sustainability, sometimes very forcefully. Las Vegas has prohibited new front lawns, limited the size of back ones, and offers people two dollars a square foot to tear existing ones up and replace them with desert plants. Between 2002 and 2006, the Vegas metro area actually managed to reduce its total consumption of water by around 20 percent, even though its population had increased substantially. Albuquerque too has cut its water use. But every water manager also knows that, as one puts it, "at some point, growth is going to catch up to you."

Looking for new long-term sources of supply, many water managers turn their lonely eyes to the Pacific, or to deep, briny aquifers that had always seemed unusable. Last August, El Paso inaugurated a new desalination plant that will allow the city to tap one such aquifer. The same month, the Bureau of Reclamation opened a new research center devoted to desalination in Ala-mogordo, New Mexico. The cost of desalination has dropped dramatically—it's now around four dollars

Southwest like Scottsdale grew by the millions during half a century of above-average rainfall. But with no end to the present drying trend in sight, the region faces an uncertain future.

per thousand gallons, or as little as $1,200 per acre-foot—but that is still considerably more than the 50 cents per acre-foot that the Bureau of Reclamation charges municipal utilities for water from Lake Mead, or the zero dollars it charges irrigation districts. The environmental impacts of desalination are also uncertain—there is always a concentrated brine to be disposed of. Nevertheless, a large desalination plant is being planned in San Diego County. In Las Vegas, Mulroy envisions one day paying for such a plant on the coast of California or Mexico, in exchange for a portion of either's share of the water in Lake Mead. "The problem is, if there's nothing in Lake Mead, there's nothing to exchange," she says.

A more obvious solution for cities facing shortages is to buy irrigation water from farmers. In 2003 the Imperial Irrigation District was pressured into selling 200,000 of its three million acre-feet of Colorado water to San Diego, as part of an overall deal to get California to stop exceeding its allotment. San Diego paid nearly $300 per acre-foot for water that the farmers in the Imperial Valley get virtually for free. The government favors such market mechanisms, says the Bureau of Reclamation's Terry Fulp, "so people who really want the water get it." At that price, the irrigation water in the Imperial Valley is worth nearly as much as its entire agricul-tural revenue, which is around a billion dollars a year. But not everyone favors drying up farms so that more water will be available for subdivisions. The valley is one of the poorest regions in California, yet the richest farmers stand to benefit most from the sale. Many more people fear the loss of jobs and, ultimately, of a whole way of life.

The West was built by dreamers. The men who conceived Hoover Dam were, in the words beneath a flagpole on the Nevada side, "inspired by a vision of lonely lands made fruitful." As the climate that underpinned that expansive vision vanishes, the vision needed to replace it has not yet emerged. In a drying climate, the human ecosystems established in a wetter one will have to change—die and be replaced by new ones. The people in the Southwest face the same uncertain future, the same question, as their forests: What happens to the stuff that's there now?

In the second half of the 13th century, as a drying trend set in, people who had lived for centuries at Mesa Verde moved down off the mesa into the canyons. They built villages around water sources, under overhangs high up in the walls of the cliffs, and climbed back up the cliffs to farm; their handholds in the rock are still visible. Some of the villages were fortified, because apparently their position on a cliff face was not defense enough. Those cliff dwellings, abandoned now for seven centuries but still intact and eerily beautiful, are what attract so many visitors today. But they are certainly not the product of an expansive, outward-looking civilization. They are the product of a civilization in a crouch, waiting to get hit again. In that period, the inhabitants of the Mesa Verde region began carving petroglyphs suggesting violent conflict between men armed with shields, bows and arrows, and clubs. And then, in the last two or three decades of the century, right when the tree rings record one of the most severe droughts in the region, the people left. They never came back.

Discussion Questions:

- What has tree ring data told us about what we consider to be normal precipitation levels in the Southwest? What implications does this have for current multi-state water use agreements?

- Describe what is expected to happen to Southwest precipitation levels over the next few decades and how this may affect the ecology of the region. What roles do Hadley cells and La Niñas play in these projections?

- Local governments have begun to implement mandatory restrictions like prohibiting new front lawns or offering incentives to tear out existing ones and replace them with desert plants. Do you think local governments should encourage, or even demand, planting drought-resistant plants or limiting residential water use for things like watering lawns or washing cars?

- Technological solutions like desalinization plants and wastewater recycling programs can help increase water availability. Are these projects a good solution, or do they have too many technical, economic or public acceptance problems to be feasible to combat the increasing water shortage problems?

- The article describes situations where San Diego pays nearly $300 per acre-foot for water that Imperial Valley farmers get virtually for free. Do you think that this is a good use of a market mechanism, or is it an inefficient and inequitable use of a scarce resource?

Map Literacy

Use the map and figures contained in this article to explain the following.

- Which areas of the Southwest have recently seen the biggest decline in rain and snowfall? What percent of the region is wetter or drier when compared with the previous 30 years?

- How does the "Percent of the West in drought" during the Medieval Warm Period compare to the recent drought that the Southwest has been experiencing? Use the associated graphs to help describe how population levels have shifted during the last century and explain how changes in these levels have compounded the problem.